No Foreign Food

The American Diet in Time and Place

Richard Pillsbury

Westview Press
A Member of the Perseus Books Group

Geographies of the Imagination

Published in 1998 in the United States of America by Westview Press, 5500 Central Avenue, Boulder, Colorado 80301-2877, and in the United Kingdom by Westview Press, 12 Hid's Copse Road, Cumnor Hill, Oxford OX2 9JJ

Library of Congress Cataloging-in-Publication Data
Pillsbury, Richard.
No foreign food : the American diet in time and place / Richard Pillsbury.
p. cm. — (Geographies of the imagination)
Includes bibliographical references (p.) and index.
ISBN 0-8133-2738-5. — ISBN 0-8133-2739-3 (pbk.)
1. Diet—United States—History. 2. Food habits—United States—History. I. Title. II. Series.
TX360.U6P55 1998
394.1'0973—dc21 97-47342
CIP

The paper used in this publication meets the requirements of the American National Standard for Permanence of Paper for Printed Library Materials Z39.48-1984.

10 9 8 7 6 5 4 3 2

Contents

Photo Credits

Illustrations

Boxes

Maps

Tables

Figures

Concept and Content: An Introduction

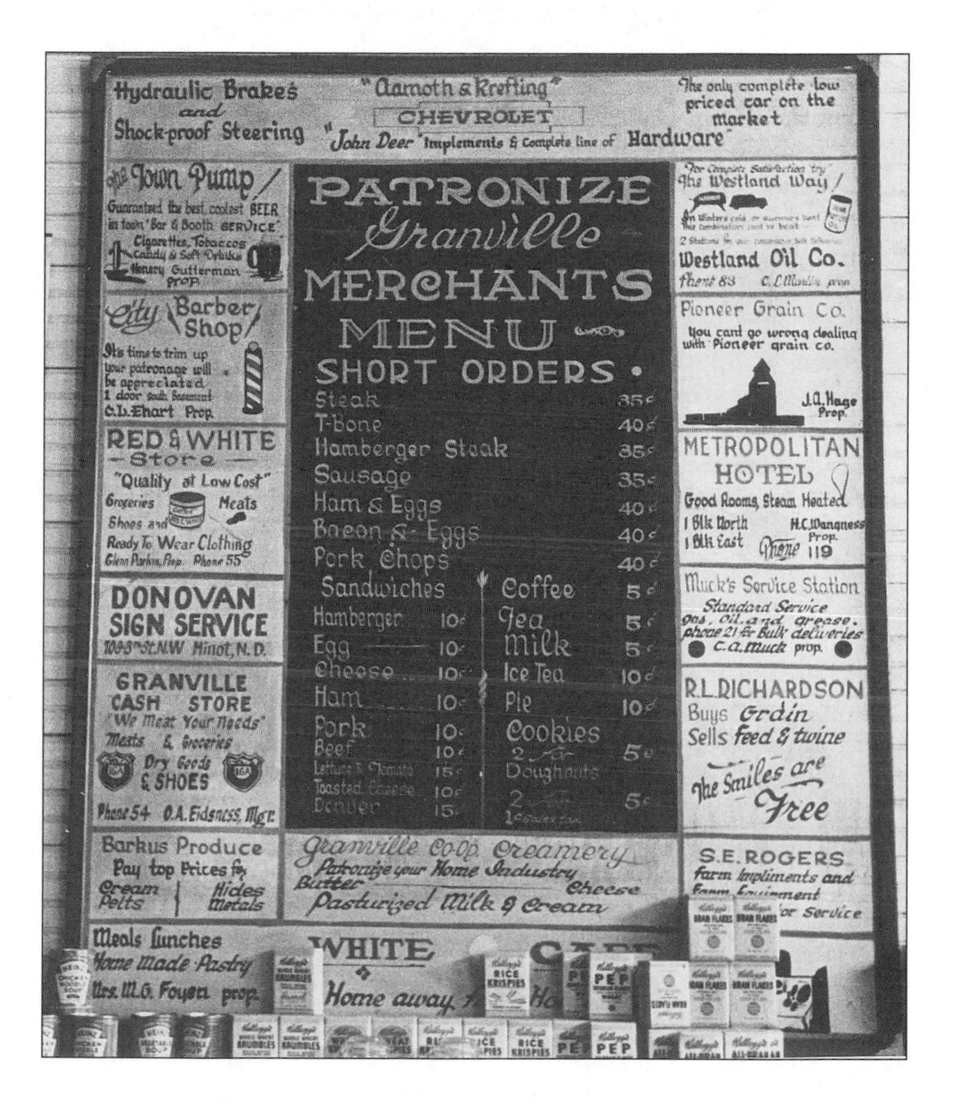

The tall, lanky waitress walked over to the table against the wall, asked the new customer for a drink order, and informed him that spaghetti was the special of the day.

The diner ordered sweet tea and looked over the tired menu and then the waitress, saying, "I just don't eat any of those strange things. I like to stick to American foods like collards and black-eyed peas."

"You're right," the waitress commiserated. "I don't eat no foreign food either."

The American way of eating is composed equally of content, the foods that we eat, and concept, the reasons we have come to eat them. Consumption of food is certainly necessary for the survival of our body, but bodily survival ultimately has little to do with what and how and when we eat. The American way of eating has been evolving for almost 400 years. The attempt to arrive at a national set of norms—a national cuisine—is at once both a stab at the obvious and a fool's errand—a stab at the obvious because our foodways are so strikingly different from those of any other culture in the world that everyone but us knows what they are, and a fool's errand because those "norms" become extremely complex when examined in detail.

Determining the American diet requires an understanding of our individual preferences and how those individual decisions fit into the whole. Each of us has idiosyncratic ways of doing things; each of us is a member of a larger group that has its own ways of doing things. Unraveling this interrelationship is difficult at best yet is worthy of the effort because in the search for understanding we come to know ourselves better.

The goal of the following discussion is to explore the national and regional foodways of America in order to gain some perspective in this rapidly changing culture. Like most Americans, I was raised believing that I was not only an individual but also a citizen of several regions—a citizen of Chico, of California, of the West, and of the nation. So much of that regional identity is breaking apart today. It is becoming more and more difficult to understand where America begins and the rest of the world ends; McDonald's and Levi jeans, CNN and the *Dukes of Hazard*, seem to have invaded every corner of the globe. The questions about these changes are endless. Does the South still exist? Where is it and why does it continue when so many other places seemingly have disappeared in a haze of MTV, Coca-Colas, and shopping malls. There will be no definitive answers here, but it is hoped that the reader will come away with a clearer recognition of how we as individuals relate to the whole. Unfortunately, each of us will leave with a slightly different understanding of these processes and images because none of us can ever become completely objective about ourselves and our ways of life.

The importance of our individuality can never be discounted. Our unique food preferences are the end product of a lifetime of personal experiences and an eternity of genetic and cultural evolution. Let us examine the food preferences and tastes of a single individual for a moment to illustrate the point. My wife, Patricia, exhibits many classic American traits. Born in western Pennsylvania, she is of mixed traditional Pennsylvanian (Scotch-Irish and Pennsylvania Dutch) and Italian stock. Many of her teen years were spent on a farm in central New Jersey where she developed some strong rural attitudes about food. Later she spent varying portions of her adult life residing in North Carolina, Alabama, Georgia, Minnesota, West Virginia, and California. Her dietary preferences represent her history in a myriad of intertwined ways. Her basic cooking style continues to be a traditional heavy Pennsylvanian approach with the understanding that no week may pass without at least one night of red sauce Italian pasta. Many of her continuing recipes, however, have been picked up along the way and represent her individual sojourns in various parts of the nation.

Typically, after raising a family while working, she would just as soon not cook daily meals at all, and we eat at least three-quarters of all our meals away from home, especially breakfast. We occasionally elect to have coffee and bagels on the screen porch overlooking the woods behind the house, but more typically the choice is to head to one of two restaurants where the coffee is hot, endless, and someone else is on the other end of the pot. In those places we find a few moments to talk without the phone ringing, allowing us to focus on the concerns of the day. In contradistinction to this lack of interest in daily meal preparation, she will spend at least a week planning Thanksgiving, New Year's, and Easter dinners and as long as a month working up the menu for our annual Christmas party.

Stir-fry is our most common home-cooked meal other than pasta. We grill almost all meat (most often chicken and, on the rare occasion, fish). Earlier *Sunset* and more recently *Southern Living* and *Better Homes and Gardens* have played a major role in reshaping my wife's cooking style. Patricia constantly examines the stream of new recipes appearing in the newspaper and magazines, and a few get tested for a while before passing out of use. However these foods are cooked, the basic ingredients have remained pretty much the same since childhood. Only butter, no margarine, too much salt, and a preference for Italian spices set the stage. Her vegetable and fruit preferences have also remained fairly constant, though the frequency of consumption of all fresh produce has increased and the quantity of red meat has decreased.

There was a time when the children were teenagers (an unappreciative audience at best) that I took on the household cooking chores and the diet perceptually changed to my preferences. A quasi-California style of cooking came into play for a few years, but ultimately I too became bored with cooking every day, the children left home, and there was little incentive to cook unless one wanted something prepared exactly to taste. My short tenure and not dissimilar tastes meant that ultimately my cooking had little impact on Patricia's tastes and preferences.

The food we order when eating out is typically the same food Patricia would have prepared at home if she weren't too busy to shop or too tired to clean up afterward. Dining at fine restaurants is random, not necessarily on weekends, and usually is oriented either toward Italian food or some entirely new experience. Our last fine-dining excursion was at Carolina's in Charleston and consisted of a glass of white zinfandel, an exquisite crab cake appetizer, a sweet potato–crusted flounder filet on a bed of stir-fried vegetables, and coffee.

Ultimately this is a picture of a typical upper-middle-class suburbanite living in a growth city almost anywhere in the nation. There is nothing in the weekly menu that reflects the southern location of my wife's residence except for a fondness for biscuits at breakfast, and her ancestral heritage appears only through a particular fondness for southern Italian cooking. Her contemporary eating habits are far removed from those of her childhood; yet those inherited foodways continue in the background to create biases and guide choices.

One's place in America today is thus a composite of one's past and one's community. The cuisine of each community has a history that has been shaped by the people who have lived there. Innovations spread rapidly in communities and regions undergoing constant growth, working their way not only down the urban hierarchy but outward across the nation. Large communities are hotbeds of change; change takes place more slowly in smaller ones, which are isolated from this tumult created by new residents with new ideas. Some areas within the larger communities too experience slower change than average because most of those living in them are little interested in change. The residents of other sections experience rapid change, especially those where corporate gypsies and other footloose Americans tend to congregate. More than two-thirds of all adult Americans do not live in the county of their birth, almost half do not live in the state of their birth, and almost 10 percent were born in a different country. This layer upon layer of individual experiences has not destroyed the preexisting regions as much as it has created a new complex map of lifestyles. Neither the map of traditional food regions nor the one of contemporary food preferences is an attempt to suggest that the areas depicted represent homogeneous sets of foodways; rather they give spatial expression to stews of preferences, some sharing common broths but varying as to a few added ingredients and others made up of a unique blend of flavors (see Maps 2.1 and 10.2). All of these stews are in the bigger kettle of expanding modernity and contracting traditionalism.

The geography of American culture has changed more since the late 1950s than at any other time in the history of the nation. The closing of the American frontier in 1890 did not keep the nation from changing. Ultimately the traditional cultural landscapes along the eastern seaboard were submerged in an avalanche of new immigrants and new economic systems. Terms such as New England, the Middle Atlantic, and even the South have less and less utility to explain contemporary America. The goal of the following discussion is to look at a single element of daily life, food, and use it to obtain some insight into the changing character of our society. The structure of the presentation is simple. After an initial examination of the factors that influence our dietary choices, a baseline for the American diet is established to help understand the processes taking place. This is followed by a lengthy discussion of the forces that are believed to have changed our diet: new transportation and storage technologies; the nationalization of the processing, distribution, and retailing industries; the development of a national media; the changing flows of immigration; and the roles of changing attitudes about and understanding of food and nutrition. A new American diet is then described in context of the forces that altered it from the traditional diet. The discussion ends with an examination of the new geography of the American diet (and culture), emphasizing the dynamics of the processes that are continuing to shape it and us.

The conversation quoted at the beginning of this introduction was overheard a few years ago in a small restaurant in North Carolina. The exchange between waitress and patron left me ruminating about American, Americanized, and "foreign" food. Obviously the entire American diet is imported, yet after a very short acculturalization period little of it is foreign. The ability of our culture to assimilate new concepts from other places is astounding and has played an important role in its evolution.

The idea that the study of food might help me understand the American scene began coming into focus more than twenty years ago. The path to this point has been tortuous. The writings of Fred Simoons and Sam Hilliard were crucial in the process. Writing *From Boarding House to Bistro: The American Restaurant Then and Now* in 1990 clarified some concepts and raised even more questions. More important, the book opened doors that I had never known existed. John Hamburger of the Restaurant Finance Corporation in Minneapolis liked my writings on restaurants and, by continually inviting me to speak at his conventions, gave me access to a large number of chain-restaurant owners. Their insights into changing American preferences provided new perspectives that helped shape a new vision of the restaurant scene. The students in my summer American Landscapes seminars also helped shape my thoughts, as they continually asked embarrassing questions about the class readings, my rambling ruminations that pass as lectures, the cultural landscape, and the real world outside the vans on our annual field trips across the South.

This book came into being when Stan Brunn of the University of Kentucky twisted my arm to extend my earlier work on American restaurants for a proposed series on the geography of imagination. This book would never have been written without George Demko and Stan's encouragement. In the same vein, the gentle editorial assistance of John Florin, and the production adjustments of Melanie Stafford and the outstanding copy editing of Diane Hess of Westview Press, gave me faith in my vision of what I thought a cultural geography ought to be and played a crucial role in bringing definition to the final product. Diane's outstanding work went well beyond copy editing, and her deft hand played an important role in the flow of the manuscript. Finally, the cartography is the always fine work of Jeff McMichael.

This is a very personal tour of American cuisine, and I apologize to the many people I have drawn into the story, mostly without their knowledge. I hope they understand that in including them I pay homage to their contribution rather than attempt to intrude into their privacy. I would especially ask forgiveness of my family and friends who have put up with my endless questions and explorations, especially John Florin of the University of North Carolina, who listened to my rambling contemplations about the American landscape for more than thirty years as we traveled the nation from Florin, California, to Pillsbury, North Dakota. Finally, I am indebted most especially to my wife, Patricia Pillsbury, a very private person who indulgently has allowed me to write of some of her private moments and sent me on my way across America, though generally believing that this effort was just the crazed musings of an eccentric and often misguided academic.

1

Concept: On What We Eat and What We Don't

And speaking of rice, I was sixteen years old before I knew that everyone didn't eat rice every day.

—Smart-Grosvenor, 1992

Every Saturday night for the first twenty-two years of my life at precisely 6:00 P.M. I sat down for a dinner of baked beans, Parker House rolls, coleslaw, and fruit pie at my grandparents' home with my entire extended family, who lived within thirty miles of Chico, California. The menu never varied, though periodically the cast did; two of my father's siblings, who lived far from this isolated community, would come when they could find time in their busy lives. The New England heritage of the family was well reflected in this unchanging menu; yet no member of my father's immediate clan had lived in New England proper for almost a century. Even the great earthenware bean pot that had been acquired in Nebraska prior to my grandfather's turn-of-the-century move to California was not from New England.

This continuing dining experience was not unique in those days, as family played a far more important role in daily American life. We may have been a bit extreme in the expectation that all the siblings would appear each week with their children and spouses, but what else would they have done in those days before television in small-town California? The pattern came to an end, of course, in the decades after World War II; my grandparents became too old to host such a large gathering (fifteen to twenty most Saturdays), and my cousins and I began leaving first for college and then for jobs scattered throughout the United States. Even the great orgiastic Thanksgiving dinners of thirty-five or more eventually disappeared as we became too scattered to even consider gathering in such an isolated place so far from the mainstream of our lives. Indeed, I have not had a single baked-bean dinner since leaving home; nor do I know what became of that great pot after my grandmother died. I moved into another world far removed from that traditional, New England farm-family milieu in which I was raised, as did all of my cousins.

All of our diets are constantly changing and are shaped not only by our past but by our daily lives. What we eat today is dependent on the way we were raised, where we were raised, and when we were raised and on the communities in which we have lived. Today I live in a community that is blessed with a grocer who has taken it upon himself to provide the greatest selection of perishable foods imaginable at reasonable prices. On any day of the year I can walk into his warehouse-sized food emporium and purchase fresh peaches. In early summer they will be local southern fruit, but as the season progresses they will be from California and even later in the season from New Jersey or Pennsylvania. It is winter now, and those sitting in bins will be from Chile and other Southern Hemisphere growing areas.

The array of food in this establishment is beyond expectation: fresh ginger from the Fiji Islands; gobee root from Japan; oranges from Israel and Spain; free-range chickens from California; fresh oysters from the Gulf Coast, Washington, and Japan; farm-raised salmon from Chile and Norway, wild salmon from Alaska, and smoked salmon from Scotland. The bakery bins are filled with more than thirty varieties of breads ranging from basic American white to Afghan flat bread, Jewish rye, "San Francisco" sourdough, and three kinds of still-warm French baguettes. An agricultural cornucopia of unparalleled variety is immediately available to me, yet when I look at my weekly menu, it is little different from that steady diet my mother served when I was a child living in a small town. The variations that have crept into my home diet stem not from the vast array of fresh foods that I can purchase at any time but rather from the give-and-take of a marriage in which each of the individuals brings a different heritage to the dining table. Thus my Italian-heritage wife thinks of manicotti and spaghetti when she seeks comfort food, not baked beans. Our menu of preferred and frequently prepared foods is an amalgamation of our individual preferences, experiences, and traditions.

All of our individual diets are constantly changing packages of preferences, availabilities, and experiences. For example, I can remember my sister-in-law from Tucson introducing the first taco to the family table and my first (restaurant) pizza—served, of course, in a reputedly rat-invested hole-in-the-wall in the wrong part of town. Tracing individual and regional foodways is a difficult undertaking; it is important to first gain some understanding of the origins of the nation's foodways.

Americans and Nutrition

Dietary habits are one of the most conservative elements of culture and one of the most difficult to trace. We know surprisingly little about the factors that shape a diet or a particular cuisine. The notion that we eat what is available is simplistic. The drive toward meeting the human body's basic nutritional needs also clearly has played only a small role in molding human dietary habits. Most nineteenth-century Americans believed that a healthy diet was an ample diet. In those times a healthy person was one who consumed vast quantities of food. The ideal human figure took on increasing girth. Likewise, one's food choices were generally felt to be of little consequence outside the ranks of a few fanatics such as Sylvester Graham and W. K. Kellogg. It was widely believed that if you liked fatback and ice cream, for example, you ought to consume them in unlimited quantities. It was volume, not content, that promoted a healthy body. The health fanatics did make inroads into American foodways, but they had no firm knowledge of chemistry, which underlies today's concepts of good nutrition.

Vitamins were not even conceptualized until a 1908 rat-nutrition research study demonstrated that unknown food elements were essential for life and health. Vitamin B was first isolated in 1911, though it was not until 1916 that it was proven that beriberi was caused by a vitamin B deficiency. The United States Food Administration began distributing literature to American schools, encouraging the consumption of milk and leaves of plants and other vegetative products in 1918 even though the actual amounts needed for good nutrition were not known.

Many vitamin-deficiency diseases were endemic to the nation prior to World War II, though their causes were unknown at the time. Pellagra, a vitamin B_3 deficiency, ravaged poor southern children for more than a century because it was not recognized that a diet based on cornmeal was vitamin-deficient. Pellagra was not stamped out until the 1940s, when better times allowed even poor southerners to broaden their diets. Even scurvy, long known to be kept at bay by the consumption of citrus fruits, was not recognized as a vitamin deficiency until 1928.

The federal government still seems undecided on exactly what constitutes a proper diet, and most Americans still have little more than folk wisdom and the half-truths propagated by advertisements to guide their nutritional decisions. The continuing popularity of various high-sugar children's breakfast cereals, french fries, and at least one brand of commercial pizza—which was heavily advertised for a time and which contains almost 1,000 calories and 30 grams of fat per slice—suggests how little progress has been made in educating the public on the virtues of a healthy diet.

Some Factors in Food Choices

It was once suggested that there are two basic factors governing human diet: (1) humans eat what they can find from their environment; (2) given a choice, they eat what their ancestors ate. The underlying truth of these statements is irrefutable, yet they give little insight into the specific factors that have shaped our diets. Only those cultures living in the most miserly of environments actually consume everything available. There is always a selection process. Some items are preferred over others; some are consumed only out of necessity. If a society's situation improves and its food supply expands, consumption of preferred foods increases at the expense of less favored items. A society practices such selection without a thought or a backward look. A complete understanding of that selection process is obviously impossible, but some factors underlying it can be determined. Some of these factors are important to varying degrees in different contexts.

Environment and Technology

The related factors of environment and technology are obviously the two overarching factors controlling what anyone eats. A food cannot become a significant part of a society's diet if it cannot be produced within the physical constraints of the local environment. These physical constraints are sometimes altered by technology; for example, new strains of food crops can increase the physical range of food plants and animals, as can irrigation and other technological innovations that alter the environment. A society can also broaden its diet beyond what can be produced in the local environment if technology is adequate to transport food in an edible condition at a reasonable cost. Innovations in transportation and storage technologies have allowed for the distribution of food to ever larger areas. Within the American context, the development of the refrigerator railcar in the late nineteenth century and produce that could be mechanically harvested in the twentieth century reduced production costs to allow the economical shipment of perishables to more areas. Today Americans take for granted that tomatoes, strawberries, oranges, and iceberg lettuce will be sitting on their grocer's shelf every time they enter the store. Environment and technology thus are so immutably linked that they cannot be separated.

Inertia

Inertia, or more simply, tradition, is an element in every diet. All people include some foods in their diets for no other reason than they have always done so. The traditional American breakfast of eggs, bacon or sausage, toast slathered with butter, and hash browns is a nutritionist's nightmare of cholesterol and other fats. Despite the availability of a variety of other foods that have been deemed acceptable for breakfast, this traditional meal continues, especially in the restaurant environment, where it is the best-selling meal

The gathering of the extended family for Thanksgiving dinner is disappearing from millions of households as work and social obligations take increasing numbers of families away from their homes on once important holidays. (Richard Pillsbury)

during the morning hours. We may say that choosing to eat such a breakfast is easier than looking for alternatives, which is to say that we consume that meal and others because they make us comfortable in their continuity.

Because we eat for many reasons other than fueling the body, the selection of foods, their times of consumption, and the combinations in which they are consumed reflect tradition as much as need and availability. This can most easily be seen in times of stress, when all of us tend to return to our own particular comfort foods. The military, for example, attempts to provide a "traditional" Thanksgiving meal to the troops in the field in order to raise morale (and probably also to effect a renewed identification with the culture they are protecting, even though many may not have taken part in that culture to the extent of eating this traditional meal in the homes they left behind). The corollary among travelers to other countries is the search for hamburgers, french fries, and pizza even if those foods are not frequently consumed by them when at home. The comfort provided by familiar foods somehow makes it easier for them to cope with sometimes overwhelmingly unfamiliar places. Conversely, foreign locals patronize expatriate American restaurants in the belief that the cachet of a Big Mac will somehow make them more akin to America and Americans. The concepts of inertia, or tradition, and comfort are immutably linked.

Unfamiliarity

Unfamiliarity also affects the acceptability of food items. Most people are reluctant to try unfamiliar foods, often citing individual tastes as an excuse. But since a small, safe taste would often quickly settle the matter, the consumer may actually be in fear of these items. How many children are exhorted to taste a new food (which they have decided is bad on sight) and even after a taste continue to protest? In my case, I detested abalone, a sea snail that has now virtually disappeared from fish markets and seafood restaurants. When I was a child, my family would journey to the Mendocino coast on the dates of exceptionally low tides to pry these ugly and unappetizing creatures from the tidal rocks. Older and wiser, I have come to covet this now expensive food; the last whole one I saw, in a fish market several years ago, was priced at $70, and it was a small specimen. Recently fish farmers have created the technology for producing these snails in pens, but they remain expensive—as well as ugly. One wonders if the public will lose its taste for this exotic dish as it becomes increasingly accessible, losing its mystique.

Acquiring a taste for a certain food thus is actually a process of transforming an unfamiliar flavor into a familiar one. Some classic foods for which Americans must first acquire a taste include coffee, asparagus, and scotch whiskey. Few Americans actually like the flavor of coffee at their first introduction. In fact, many must add large quantities of adulterants—sugar, cream, and, more recently, assorted spices—to make this drink tolerable; yet this beverage has become the nation's most identifiable national drink. One must wonder how many of those millions of consumers like the drink and how many consume it because of social convention or to obtain a cheap caffeine fix.

An extension of this thought is that societies universally reject foods that disgust them in some way. Our society feels disgust for insects and snakes—for virtually any crawling or squiggling creature—and for almost anything else that appears strange. This disgust has little to do with the actual taste of the item. Rattlesnake meat, for example, is almost universally reviled—some vomit when they discover what they have just consumed—yet many who try it actually find it agreeable.

Overfamiliarity

In his pace-setting study of food in the 1960s, Fred Simoons demonstrated that the history of most food avoidances is far more complex than it appears. His critical examination of the Western European rejection of horse flesh, for example, indicated that Americans' disdain for this food generally does not relate to an association with a famous horse such as Trigger or a beloved horse in one's everyday life (overfamiliarity) but rather is specifically attributable to an A.D. 732 decree by Pope Gregory III that Christians should not eat horse. The pope's reason for this decree did not relate to the horse as an animal at all; rather it stemmed from the church's desire to control the heathen in Germany and northern Europe who consumed horse as a part of their pre-Christian ritual. Believing that some northern European Christians had incorporated ancient heathen beliefs into their Roman Catholic rituals, church leaders hoped to either stop the practice or be able to identify the malefactors. This strategy did not work, and horse is still consumed—without religious connotations—in those same portions of northern France, southern Germany, and the Low Countries. Indeed, the large-scale exportation of horse meat for human consumption in Europe is becoming a cause célèbre for increasing numbers of animal rights activists. The continuing modern American avoidance of horse meat probably stems less from the medieval ban than it does from a general drift away from the consumption of animals not a part of the American mainstream, including the nontraditional squirrel, rabbit, and buffalo. Indeed, the attempt to introduce buffalo flesh into the American diet because of its reputed positive qualities is an interesting example of Americans' general resistance to consuming the unfamiliar even if it is better for them.

Social Status

Social status is also an important factor in food selection. Virtually all food has some association in this regard. Some believe that by consuming high-status foods, they are able to raise their social standing. Filet mignon, beluga caviar, and lobster carry high-status social cachets in our society. Red beans and rice, chicken necks, and catfish are often presumed to be consumed by people of low status. As my mother said derisively when my father, having read about catfish in the South, decided to catch some in the Sacramento River for the dinner table, "I hope you don't expect me to cook those things. Only Okies eat catfish."

Traditional societies controlled the consumption of high-status foods by forbidding their consumption by those deemed unworthy for reasons of gender, age, and family or clan. Thus pregnant women might be prohibited from eating pork because of the humanoid look of the pig fetus, and the bear clan might be restricted from eating that animal because of its association with the clan's heritage. Our society uses price as a controlling factor, and high-status items often carry higher prices than their mere "quality" or cost of production warrant. The theoretical egalitarianism of American life is reflected in the nation's use of economics to control food prohibitions based on class: Anyone in the society can indulge in such foods if he or she is willing to make the financial sacrifice. Among those who are willing (and able) to pay is the young swain who wants to impress his date on prom night. A familiar sight during the American prom season is the startled look on a young man's face when he is presented with the whole lobster he ordered to demonstrate his worldliness. He has no idea which part to eat, much less how to attack this strange crustacean stretched across his plate. His concern soon turns to panic as he wonders how to salvage at least a bit of dignity. All this happens while the waiter, with a bit of a smirk, is tying the obligatory bib around the lad's neck and offering him a strange fork.

Perceived Properties

The perceived properties of foods are an inherent element of all food consumption. From Neanderthal to modern suburbanite, perceptions rather than actual knowledge have determined the foods we eat. Many foods are eaten because we *believe* they impart desired qualities to us. How many children in America, a society that generally does not consume organ meats, have virtually been force-fed fried liver because it was "good for them"? How many came to eat canned spinach during the 1930s and 1940s after watching Popeye cartoons, which extolled the strength-giving qualities of this food? Similarly, we eat a host of other foods because we believe that their consumption will enhance our lives—carrots to see better, lettuce for its vitamins, and beef to make us strong. How different are these beliefs from those of other peoples who consumed deer to make them swift, bear to make them strong, or even the heart of the enemy killed in battle to make them brave?

It is fashionable to deride these supposedly primitive beliefs; yet a popular book in the 1980s proclaimed, "Real men don't eat quiche." Although the trendy admonition was certainly made tongue-in-cheek, no doubt some men stopped consuming this then chic restaurant entrée in fear that they would be perceived to be deficient in manliness. Quiche has largely become passé today and less frequently appears on menus; nutritionally, this is probably a positive development, although the high cholesterol of some forms of this dish had little to do with its demise. Nevertheless, refusing to eat it was for a time and within certain social groups a de facto symbol of American manhood. Consumption of specific foods is tied to concern about how we are perceived by our peer group.

Fealty or Group Membership

Fealty or group association is another important factor in the evolution of food preferences and avoidances. As previously suggested, many in our society consume high-status foods as one method of being perceived as a member of a particular group; in actuality, this concept is much broader than status. It has been argued that one of the most important elements in pork avoidance among Muslims was the pig's association with urbanites, who were generally perceived as dirty and uncivilized by the nomadic tribes of the Middle East. It has been suggested that Muslims traditionally consumed camel to demonstrate their association with their group and avoided alcohol and pork to demonstrate their faith. Similar behaviors associated with food and group identity are common in American society and include going out to get a beer with the boys after work.

The prescribed menus for most holidays also play an important role in demonstrating group membership. All-American turkey is preferred at Thanksgiving (a nationalistic holiday), but ham is preferred at Easter (a preference that clearly delineates Easter as a Christian holiday, as pork is banned from the Jewish diet). Holiday meals are an important way for individuals to demonstrate their cultural identity with a group. In contemporary America, Thanksgiving turkey with (corn)bread stuffing, mashed potatoes, yams, pumpkin pie, and cranberry sauce is de rigueur. We believe that by consuming these items we are affirming our ancestral roots and celebrating national honor and familial solidarity. We try to include as many Americans as possible in this ceremonial meal, whatever the cost. The military provides even troops in battle with the proper holiday foods (as duly shown on the 6 P.M. evening news) and local communities reach out to often-underfunded urban soup kitchens to provide the nation's most downtrodden with the proper celebratory meal (also featured on TV news). The fact that we are unsure of what the pilgrims ate during that November feast and that the Thanksgiving celebration and accompanying meal actually represent a gross commercialization of an event memorialized by a society searching for a common identity that

Thanksgiving: Expressing Group Identity

Thanksgiving became an official national holiday only after Sarah Josepha Hale, then editor of *Godey's Lady's Book*, essentially browbeat President Lincoln into proclaiming it as a national holiday during the dark days of the Civil War in 1863. The South, which did not celebrate the holiday, was not in a position to complain, and through time this celebration of the nation's supposed New England roots became increasingly important. Maintained by annual proclamation for almost eighty years, Thanksgiving received the status of a legal national holiday at the beginning of another cataclysmic conflict in 1941.

The standard holiday menu seems to have been largely invented by cookbook writers and home magazines with a little help from Norman Rockwell's now iconistic painting. The menu certainly is an idealized vision of that first seventeenth-century meal. All that is actually known about the feast is included in a firsthand account by Edward Winslow: "Our harvest being gotten in, our governor sent four men fowling, so that we might after a special manner rejoice together after we had gathered the fruit of our labors. They four in one day killed as much fowl as, with a little help beside, served the company for a week. . . . Many of the Indians coming amongst us, and among the rest their greatest King Massasoit, with some ninety men, whom for three days we entertained and feasted, and they went out and killed five deer, which they brought to the plantation and bestowed on our governor, and upon the captain and others. And although it be not always so plentiful as it was this time with us, yet by the goodness of God, we are so far from want that we often wish you partakers of plenty" (Bradford and Winslow, 1969).

Precious little is known about what was actually served at that feast. If turkey was one of the fowl served, it was a far different bird than now graces modern tables. The domesticated turkey was actually developed in nineteenth-century Europe from imported American birds. The new reengineered domestic bird was not reintroduced to America until the mid-nineteenth century. Reintroduction, however, did not bring commercial success. Commercial turkey production was so unimportant in the nineteenth century that flock counts were not included in the general poultry category in the census of agriculture. Large-scale commercial production did not come into existence until the late 1920s. Though the turkey was painted as an idealized full-breasted bird in Rockwell's interpretation of the gathering, today's full-breasted bird was largely created by modern genetic engineering after his painting was completed. The modern bird is so grossly top-heavy that some of the creatures can barely walk when they are sent to slaughter.

(continues)

(continued)

Cranberries were indigenous to New England but probably were not a part of the first Thanksgiving meal. Commercial cranberry production and canning did not begin until the late nineteenth century, and widespread distribution is largely a mid-twentieth-century phenomenon. Cranberry production initially was concentrated in southern New England, and the Ocean Spray Company of Cape Cod continues to dominate the industry, although secondary production centers in Wisconsin and New Jersey are responsible for ever-increasing percentages of the total harvest.

The Pilgrims probably had never seen an "Irish" or white potato—they were not successfully cultivated in this country on even a moderate scale until more than a century later. The white potato did not enter the general American diet until almost a century after that. Gravy, or more specifically meat drippings, as a flavoring was known but was usually served with bread.

The various species of yams and sweet potatoes are indigenous to Southeast Asia, tropical America, and possibly West Africa. Yams were widely grown in West Africa by the sixteenth century, and it is believed that their use was introduced to America from there. The crop was unknown to the typical Englishman at this time and could not possibly have been grown in New England because of environmental constraints. Spanish potatoes (yams) were imported to New England during the eighteenth century.

Both Thanksgiving Day and the feast are largely nationalistic myths created to provide a past that never was. The very nature of the dynamic American evolution has meant that few of us have any genetic ties with that event. Yet millions of Americans dutifully trot out their turkeys and cranberry sauce on the third Thursday of each November to celebrate an event that has come to symbolize group membership—disregarding the reality that many of our ancestors probably would not have been welcome even if they had been in the vicinity at the time.

never existed is immaterial. Thanksgiving has become a Christmas-like holiday in that it involves cards and family visits, but it is open to all regardless of religion or ethnic heritage. And as with Christmas, it is becoming increasingly difficult for most Americans to achieve the idealized Thanksgiving feast depicted by Norman Rockwell on his *Saturday Evening Post* cover many decades ago. Families are widely scattered across the country, grandparents no longer live just down the street, and this kind of repast is difficult to prepare for the average 3.22-member American household.

It is not my goal to debunk the classic Thanksgiving holiday but rather to suggest that the concept of consuming selected foods to obtain their perceived qualities has not disappeared from modern society. Nor is this phenomenon restricted to national holidays. We demonstrate membership in a group every time we provide food for others, whether it be shown by participating in a church social or inviting people to our homes to break bread. What we serve speaks of our community ties with those to whom we serve it. No member of a midwestern church would dare arrive at a church supper with a plate of mussels or escargot or for that matter red beans and rice. Those things would not fit that group's perception of the continuity and temporality of its community, and by bringing such foods, you would be sending a message that you do not acknowledge your membership in and acceptance of the rules of the group.

I have attended the annual church homecoming of a small rural Baptist church in central Georgia for more than twenty-five years as a guest of an old friend. The progression of food offered by the members at this special time well reflected the community's evolving self-image as it changed from an isolated, declining, poverty-ridden cotton community with little future to a community with a growing base of exurban Atlantans. The fare at this annual event has changed as the community has changed. Early on, the tables were dominated by classic southern cuisine done to its finest turn. Modernity began creeping in as the first retired returnees from the city tried to demonstrate their worldliness. The arrival of increasing numbers of trailers on the back lots of farms and the beginnings of renovation of the stock of Victorian farmhouses brought more change in the 1970s and 1980s. More and more *Southern Living* dishes began appearing, and the traditional dishes became more and more modernized. Today, the most traditional dishes are most likely to be brought by the community's newest residents, reveling in their return to "their" roots; the most modern are provided by the housewives who never left and who are happy to be freed from the shackles of traditionalism, which held them fast to a limited diet most of their lives.

Time and Place

"Everything has its time and place" is a consistent theme in the American self-conception and certainly plays a role in the nation's foodways. The how, the why, and the where of consumption all play important roles in determining whether a food is acceptable. Certain foods are perceived to be proper at specified times. A proper American does not drink a cola and have a slice of pizza as the first meal of the day—if we overlook the penchant of teenagers to prove they can break any of society's rules at will. For many years it was virtually impossible to successfully order a soft drink in a restaurant during the breakfast period—the fountain machinery typically was not even cleaned and set up for the day. Even today McDonald's serves breakfast only until 10:30 A.M. even though customers might be standing in line to order it. Nor can one buy a hamburger before that time. We are told that there is no demand for breakfast at McDonald's after 10:30; yet breakfast items account for 58 percent of total sales of one of the nation's larger family restaurant chains and 67 percent of another.

The issue of drinking alcohol is especially intriguing in our society because there are so many rules about its consumption. Age distinctions are most prominent, but time considerations are also important. Interestingly, our society considers drinking a martini the first thing in the morning to be a sign that the consumer may be an alcohol abuser but accepts the consumption of the same amount of alcohol in tomato juice, called a Bloody Mary. The history of American attitudes about alcohol consumption is an interesting one; the nation has moved from unrestricted consumption to full prohibition and back to controlled consumption. Further complicating the history of alcohol prohibition in America is the fact that some of its supporters acted out of self-interest. Henry Ford financially supported the prohibition movement because he was trying to ensure the presence of a sober workforce at his factories. In much of the South today, control of alcohol sales is often supported by a coalition of religious leaders and bootleggers.

We have particular ideas about the acceptability of foods in terms of when they are consumed. Dessert must be consumed after a regular meal, seemingly must be explained to other diners if consumed otherwise, and is generally considered inappropriate for breakfast. But doughnuts, cinnamon rolls, and other sweet bakery items, containing essentially the same ingredients as traditional dessert items, are acceptable breakfast foods. Indeed, these items are acceptable at almost any time of the day or night either alone or in conjunction with a beverage.

Body and Soul Food

Ultimately what we eat and when we eat are the result of a complex set of decisions, most of which take place subconsciously. It has been suggested that one of the basic elements in the decision process is the goal of survival, yet this ultimately plays a surprisingly small role in either personal habits or the evolution of our national cuisine. One rarely eats solely to fill the stomach.

Obviously all food fuels the body; some foods are consumed only for that reason, and others are consumed because they elicit a desired emotional response. Almost half of the entire nation's consumption of pumpkin pie takes place during a single week each year. Its consumption is based on a set of beliefs and traditions that may or may not have validity, but whether or not these beliefs are based in reality, it is clear that the Thanksgiving pumpkin pie eaters are feeding their souls, not just fueling their bodies. One's special birthday cake similarly brings memories of an idealized childhood. My own childhood birthday cake, for example, was an applesauce cake baked in an angel food cake pan; it was heavy as a rock, created from a recipe that was repeatedly altered by a doting mother. My wife, who came on the scene after my mother's death, spent years trying to recreate this childhood memory until a favorite aunt pointed out that my mother utilized several recipes, that she never copied a recipe without altering it, and that there actually never had been a single "Dick's applesauce cake." But I still dream of that cake, which apparently never existed.

Some Final Thoughts

Food is about memories, traditions, and history. An exploration of the geography and development of American regional and national cuisines thus must include the contextual evolution of our entire dining experience—from the accessibility of the ingredients to their means of preparation to the ways and times of consumption. Understanding the evolution of our contemporary diet thus must begin with an exploration of the nation's first eating habits and its early "geography of food." A comparison of our past and present food geographies will help us understand the evolution of our foodways. The American culture and diet is undergoing rapid homogenization; yet even a hurried trip through south Houston or Seattle or Harlem suggests that definitive regional differences still exist in the nation's food preferences and consumption modes. Still, many traditional, regional cultures have disintegrated, and have been replaced with new ones. In the next chapter, a baseline of traditional foodways will be established so that we may better understand today's dynamically changing culture.

2

Content: A Traditional American Diet

I hold a family to be in a desperate way when the mother can see the bottom of the pork barrel. . . . Game's good as a relish and so is bread, but [salt] pork is the staff of life.

—James Fenimore Cooper, 1958

Colonial America was a rural land with few urban communities. More than 95 percent of the population lived as farmers. Water transport was comparatively inexpensive, but only poor roads connected interior residents to the coast. The high cost of shipping commercial produce to market forced most interior farmers to concentrate their efforts on subsistence farming through most of the eighteenth century. Livestock typically was the first commercial crop attempted by inland farmers because most of the animals could be walked to market. Thousands of cattle, hogs, and even geese were walked across Massachusetts, Pennsylvania, and Virginia to markets in Boston, Philadelphia, Baltimore, and other cities. Farmers near those cities soon realized that they could purchase those road-weary animals, fatten them for market, and sell them at a profit. Hundreds of these farmers soon were raising animal feeds and purchasing arriving animals as their primary business.

Some frontier grain farmers solved the problem of high transportation costs by selling their output to local distillers who converted the grain into high-value, low-weight whiskey. Americans were heavy drinkers, and the market for inexpensive distilled spirits was strong. Once established in the interior, the distilling industry expanded rapidly. Spying a potential source of revenue, the federal government first attempted to raise funds from sources other than import and export duties with a whiskey tax in 1793. The resulting Whiskey Rebellion brought the infant government to its knees as farmers revolted across southern New England, upstate New York, and the upper Ohio Basin. Whiskey distilling continued unabated after the matter was resolved; in an 1800 census, more than 800 stills were recorded in western Pennsylvania and northern West Virginia alone.

More and more eighteenth-century farmers near the growing urban areas began realizing that greater profits could be realized by producing specialty crops rather than low-value grains and livestock. Specialty farms soon began appearing around New York and other cities to supply those centers with vegetables, fruit, eggs, poultry, and dairy products. By 1810, farmers in Essex County, New Jersey, were producing more than 300,000 gallons of applejack a year for nearby New York and Philadelphia.

The Staple Ingredients

Americans consumed vast quantities of meat in comparison to their European counterparts throughout the nineteenth century. One 1848 estimate indicated that adult Americans consumed on the average 300 or more pounds of meat per year. Beef and pork were preferred, but almost everything from squirrel to lobster was consumed in large quantities. Salt pork was the single most important red meat consumed throughout the nineteenth century. Pigs were easy to raise, ate a variety of feed, and matured rapidly. Most farmers cured their own salt pork in the early days, but even they began to utilize some commercially packed meat as the industry evolved. Beef was also cured, but pork was the staff of life for most Americans throughout the nineteenth century.

Fresh-meat consumption was comparatively rare until after the Civil War in most areas except during the fall slaughter period. The lack of refrigeration and the fear of spoilage made most Americans afraid of uncured meat. The Department of the Army required Chicago meat packers to supply some quantities of fresh meat to the Union armies during the Civil War, and this helped instigate the move toward eating and transporting fresh meat. Fresh-meat production by the Chicago Union Stockyard packers increased after the end of the war, although total quantities continued to be relatively small until the widespread appearance of refrigeration.

Chicken-consumption levels during the colonial period are a mystery. Chickens are described as a part of almost every farmyard; recipes utilizing chicken meat are found in every early American cookbook. However, the various nineteenth-century Census of Agriculture poultry statistics suggest that chicken production was comparatively modest and that home flocks may not have been as extensive as travel accounts imply. Eggs were an important ingredient in cooking, discouraging farmers from slaughtering any chicken that did not have declining egg production, except on special occasions. When these facts are coupled with the knowledge that it took the colonial chicken more than 200 days to mature in comparison to only about 45 days today, one must conclude that the role of chicken in the daily eighteenth- and nineteenth-century diets has been overemphasized, probably because it was such a treat.

We are told that the inshore waters of the colonies teemed with fish waiting to be caught. Oyster houses began appearing in the late eighteenth century as some of the first freestanding restaurants in the nation, and there were many fish recipes in the cookbooks of the time. The state of the transportation network, however, tended to limit how far fresh fish could be transported without risking spoilage. Amelia Simmons (1796) commented that fish brought to market in panniers (baskets) were more likely to be fresher than those carried in bags because the heat of the horse speeded the deterioration process. She further suggested that salmon, "unlike almost every other fish, are ameliorated by being 3 or 4 days out of water, if kept from heat and the moon, which has much more injurious effect than the sun." In reference to shad, she commented that she had "tasted Shad thirty or forty miles from the place where caught" and that they had had a good flavor. Implied was that she considered these distances and times to be near the maximum possible to transport fish with any hope of retaining minimal quality. Wagons traveled at a speed of about two or three miles per hour, so it is difficult to imagine that most fresh fish could safely be transported more than twenty miles from where they were landed. Residents within ten miles or so of the coast thus had constant access to large quantities of lobster, oysters, mussels, cod, and other finned species, which ultimately played a great role in their daily diets. Inland dwellers in most of New England, New Jersey, and eastern New York, in contrast, could obtain only salt cod and other preserved fish; those living far from the coast in places such as western Pennsylvania, New York, and Virginia could not purchase ocean fish until the development of canal and rail systems.

Bread was the second element of the basic diet and was consumed in large quantities. Wheat bread had been preferred in Europe, but wheat and other small-grain flours were often too expensive for regular use by most colonists in New England and much of the South. Coastal residents in those areas could purchase imported grain flour relatively inexpensively almost from the beginning, but frontier subsistence farmers generally consumed some form of cornbread until sufficient quantities of wheat flour could be obtained. Wheat rust became a major problem in New England after the middle of the eighteenth century, further decreasing local supplies of wheat for flour while increasing supplies of rye and oats, which were grown in its stead. Certainly the wealthy consumed more wheat bread than the poor, but everyone consumed at least some cornbread throughout the colonial period.

The cost of wheat flour was not the only factor favoring the continuing consumption of cornbread in many areas. Brick ovens built into the fireplace were comparatively uncommon throughout much of the colonial period. Most home baking took place either in Dutch ovens or in reflector ovens placed in the fireplace until the invention of the cast-iron, wood-fired cookstove in the nineteenth century. Millions baked bread in this manner, but it certainly was easier to deal with cornmeal bread products and other nonyeast

Home production and processing of the bulk of a family's food continued well into the nineteenth century for many isolated American families. (James Mooney, Bureau of American Ethnography, Smithsonian Institution, 1888)

bread products. The paucity of ovens was also reflected in the preference for boiled or grilled meats. The transition to oven-roasted meats in the late nineteenth century was made only after great debate.

Finally, there was the problem of yeast. Emptins (a form of yeast obtained from the remains of the brewing process) were available in many areas, but reliable commercial yeast was not marketed until the 1860s, when the Fleischman brothers patented their process. Mary Randolph included a recipe for making yeast cake in her 1825 cookbook, but that was well after the colonial period. Pearl ash and several other leavening agents became known in the late eighteenth century, but knowledge of them spread slowly.

Amelia Simmons's *American Cookery* (1796), the nation's first homegrown cookbook, did not include a single recipe for yeast bread, yet it did have recipes for six kinds of rusk (a flatish, hard, baked cracker), three kinds of biscuits (which also are really crackers), and nine pastes (doughs) for tarts and pot pies. Several of the rusk recipes included emptins, but virtually all included sugar, which aids the fermentation process. Soft wheat bread as we know it today apparently was largely a nineteenth-century phenomenon in most middle- and lower-class homes in America.

Johnnycake, or journey cake, did universally appear in the cookbooks of the era. The origins of johnnycake are obscured. These breads are obviously the easiest way to transform grain meal or flour into a consumable product and have been around in varying forms since Neolithic times. Flat, baked hearth breads of oatmeal and other grains had been a staple of the British poorer classes at least since medieval times. Tortillas are conceptually the same product, as are the eastern Amerindian pone breads. The advantage of all of these breads, also called hoe cakes and ash cakes in Europe, was the simplicity of both ingredients and cooking implements. They were typically made of just meal or flour and water with a bit of salt if available. They could be baked on a flat rock, in the ashes, or on a griddle next to the fire. It was an obvious transition for the Europeans to adapt Indian (corn)meal to make this common food in America.

The remainder of the typical menu varied widely depending on the local environment. Seafood was a very important dietary element where available, as was game on the edge of settlements, where it was available in sufficient quantities to provide a significant dietary supplement. A broad range of European vegetable crops were immediately introduced to America, including turnips, parsnips, cabbage, onions, carrots, beans, and peas (a generic term here to include a variety of beans and legumes). These vegetables played an important role in the European diets but were comparatively less important in the New World, where meat was more readily available.

Soups, potages, and porridges played an important role in everyday cooking in medieval England, and they continued to be the primary dishes including vegetables in this country. A classic English recipe might have instructions to dice celery, endive, spinach, sorrel, leeks, onions, and cabbage and place them in a broth with a bit of salt pork, a fowl (not necessarily chicken), and cook. A large section of recipes for soups and potages was included in every early American cookbook. Succotash was the most widely distributed American dish of the potage type; baked beans fall into the same general class of dishes. Virtually all vegetables in the South were cooked in pots and allowed to simmer for long periods. Both gumbo and jambalaya are examples of this type of dish; the most famous (and controversial) of the southern potages was Brunswick stew.

It is difficult to ascertain the impact of native dishes besides those using cornmeal, which served as a grain flour substitute in early colonial diets. For example, travel accounts indicate that succotash (a vegetable dish of corn and lima beans) was a common dish in many areas; yet many early cookbooks do not carry recipes for this seemingly basic item. Pumpkin and some other New World squashes, a variety of beans, and game became important elements of the colonial diet, though the actual volume of consumption in, say, 1790 in central Connecticut is unknown.

A host of foods generally thought to have been common in colonial America were either little known or consumed in only a few areas. Virtually any

Brunswick Stew, Burgoo, Succotash: Regional Variations

The medieval vegetable potage did not die out in the New World during the colonial era; rather it took on an American look. Succotash, a European adaptation of a native dish, was the most widespread single recipe for this type of dish. Hundreds of variations exist of every traditional recipe in the United States. Some of these variations are expressions of individual preferences of local cooks, some stem from regional taste preferences, and some stem from the time in history that the recipe was written. The recipe for succotash shown below, for example, was a 1769 attempt to create a meal memorializing the landing of the Mayflower. The authors of the recipe apparently were unaware that white potatoes were not successfully tilled in the United States until 1719. Most early succotash recipes are based on lima beans instead of white beans; some later versions also include tomatoes.

Succotash, Old Colony Club (1769)

Boil 2 fowl in large kettle of water. Remove fowl. Simmer 2 pounds of dried white beans with 1/2 pound of salt pork. Add 4 lb. piece freshened cornmeal brisket and cook with diced turnip, 5 or 6 sliced potatoes, 4 qts. cooked dried corn. Return the meat of one fowl to mixture and continue cooking.

Brunswick County, Virginia; Brunswick County, North Carolina; and Brunswick, Georgia, all claim to be the original home of Brunswick stew, though in actuality the stew was probably first made as a variation on the even earlier succotash and may have been "independently" created several times from this origin. The most popular story is that the gentility first became aware of this dish on a hunting trip in southern Virginia in 1828, when Uncle Jimmy Mathews cooked a stew for Dr. Creed Haskins. The original stew reputedly was essentially made of squirrel and onions, but it was only a few years before it began to evolve into the popular dish served at political rallies and other large events.

An 1872 Virginia version of this dish is included in *Housekeeping in Old Virginia* and is attributed to Mrs. R. of Lynchburg. This version shows the beginning of the decline of squirrel meat in mainstream recipes; other recipes from this period use only chicken. Pork is the most common meat on the Gulf coastal plain.

Brunswick Stew (1872)

Take two chickens or three or four squirrels, let them boil in water. Cook one pint butter beans, and one quart tomatoes; cook with the meat. When done, add one dozen ears corn, one dozen large tomatoes, and one pound of butter.

(continues)

(continued)

Take out the chicken, cut it into small pieces and put back; cook until it is well done and thick enough to be eaten with a fork.

Season with pepper and salt.

(Hill, 1872)

Potatoes typically were used less and less frequently by cooks further to the south, primarily because white potatoes were rarely available more than a few months of the year. A typical recipe of the Gulf coastal plain, for example, also includes peas rather than butter beans.

Brunswick Stew (1976, Swainsboro, Georgia)

15 lb. chicken
2 lb. 8 oz. diced celery
3 lb. diced potatoes
12 oz. melted oleo
1 1/4 gal. chicken stock
3 lb. 8 oz. diced pork
3 lb. 8 oz. diced carrots
1 lb. fine chopped onions
12 oz. flour
1 lb. frozen peas

(*Pineland Country Cooking: From the Geechee to the Hoopee*, 1980)

The trend becomes even more pronounced as one looks at recipes from the western and southwestern Gulf regions.

Brunswick stew also traveled westward from Virginia but became known as burgoo. Although both the Brunswick stew and Kentucky burgoo aficionados would deny this assertion, examine the following recipe and draw your own conclusions:

Charles Patteson's Kentucky Burgoo (1988)

1 stewing chicken
6 ripe tomatoes, cut up
2 tsp. curry powder
1 tbs. coarse (Kosher) salt
2 skinless chicken breasts, cut up
2 c. fresh or frozen corn kernels
2 c. shelled fresh lima beans
1 tbs. filé powder
4 c. beef stock
2 medium onions, unpeeled
1 tbs. black pepper
1 1/2 cups bourbon
1 c. country ham (optional)
1 c. diced raw potato
2 1/2 c. okra

Patteson notes that although there is great variety in recipes, there are some ingredients in common, including tomatoes, lima beans, onions, potatoes, okra, and corn. Most also include one kind of fowl and some kind of red meat.

(continues)

(continued)

The regionalization of traditional dishes such as Brunswick stew often follows predictable trends as the recipes spread from their original home. Brunswick stew began as a squirrel stew and evolved into a largely vegetable stew of cornmeal, whole-kernel corn, lima beans, and white potatoes—essentially a southern succotash. White potatoes were rare much of the year on the muggy Gulf coastal plain and were soon replaced with increased amounts of cut corn and peas. In Texas and beyond, the dish ultimately became westernized with even more cornmeal and some peppers and lost its southern flavor altogether.

Brunswick stew and burgoo were rarely cooked at home. Burgoo is mostly associated with large groups, especially political rallies. About the only place one regularly encounters Brunswick stew is in the thousands of barbecue restaurants scattered across the South, though more and more of these places are passing off commercial versions of this distinctively regional dish. If it is served at home it will be either from a can or, more recently, frozen. Neither version is up to the standards of a good barbecue pit.

food in the world could be imported to coastal cities and was available to those who could afford it. Recipes including citrus fruit, mangoes, coconuts, most spices, Spanish potatoes (sweet potatoes), and rice routinely appeared in the cookbooks of the early nineteenth century. Those foods generally were not available to the middle and lower classes anywhere and were rarely available to even the rich in the interior. Tomatoes, a member of the nightshade family, are probably the best-known modern favorite not generally available in colonial times. The white potato also was rarely grown and consumed in America until after 1750 and was not common until the nineteenth century. Yams were widely consumed in the South but rarely served elsewhere, as was the case with peanuts, watermelon, rice, and an assortment of other dishes.

Americans have always consumed more water than most Europeans; their other preferences also took some other sharp departures. Colonial Americans drank vast quantities of alcoholic beverages by modern standards. The Rev. Jonathan Ashley of Deerfield, Connecticut, for example, included a list of his food expenses in a request for a salary increase in 1745. He allocated a full 8 percent of his gross income for assorted alcoholic beverages (Coe and Coe, 1984). It would appear that the typical eighteenth- and nineteenth-century American drank often and much.

Preference patterns for distilled beverages underwent a significant change in the early nineteenth century. The end of the molasses trade with the

Caribbean brought a comparatively rapid decline of the previously favored local rum as the price increased. The less expensive American whiskey was more than an adequate substitute and inexpensive as well. Though generally assumed to be distilled from corn, whiskey can be produced from virtually any grain. Rye whiskey was especially popular in Pennsylvania, and that state continues to be the center of the production and consumption of rye whiskey.

Ale and beer were always important even during colonial times. Easy to brew for even the individual householder, these beverages were common in all of the colonies. The arrival of increasing numbers of continental Europeans, especially after the Civil War, encouraged the development of large, commercial breweries, reducing the relative importance of homebrewed beers and ales in urban areas.

Wine was less widely consumed during the colonial period, though attempts to introduce winemaking started in the seventeenth century. These efforts failed, and with the westward migration of the center of settlement, wine became less and less important to the national diet. The economic and intellectual elite and their pretenders continued to consume large quantities of wine, but even today the nation's wine consumption is minuscule in relation to its total alcoholic beverage intake.

Seasonal Diets

Almost everyone throughout the nineteenth century followed "winter" and "summer" diets interspersed with transitional periods of increasing or decreasing fresh food. Meat curing was a seasonal activity. Farmers slaughtered their stock soon after the first freeze in the fall. The crisp mornings and cool afternoons allowed them to butcher and smoke or salt the fresh meat with a minimum of spoilage. On the farm this was the only time that most families consumed quantities of fresh beef and pork before the widespread use of refrigeration. Even the largest of the meat packers before the Civil War did not begin the packing season until after the first frost each year and quit each spring as the weather turned warm.

Fresh fruits and vegetables virtually disappeared from winter tables except for a few cool-weather vegetables like cabbage, onions, melons, and turnips that could be carried through at least part of the winter. Soon these too were gone, and only pickled and dried fruits and vegetables remained. Pickling and dehydrating were important summer activities on most farms as families worked to provide winter provisions. Cucumbers, green melons, barberries, pigs' feet, sauerkraut in Germanic areas, and eggs all went into the pickling crocks and barrels found on every farm. Large quantities of fruit were also preserved by cooking in heavy syrups and generally storing in glass. The thousands of stoneware and pottery crocks still found in every antique shop in the eastern United States stand as reminders of these activities.

The kitchen garden and its orchard of apple, peach, and other fruit trees also provided large quantities of fruits and vegetables for drying. Apples were the most widespread of the dried fruits, and recipes abound on how to make pies and other desserts directly from the dried product. Peaches and apricots were also commonly dried. Leather breeches (green beans) were probably the most common of the dried vegetables. Strung across the porch on strings, they were a common sight in rural Appalachia until after World War II, and even then, the older residents continued to prepare this old favorite.

The first signs of spring greenery brought a change in diet in rural areas. Dandelions were often the first green to appear in many areas, and "salats" of mixed wild greens were a common addition to many early spring meals. In his 1774 garden book, Thomas Jefferson listed peas as the first green picked. A bit more prosaic was the ramp (a type of wild onion), which was the first spring green in much of Appalachia. Its importance in traditional life was long commemorated in festivals in West Virginia, Kentucky, and elsewhere until the 1970s influx of urbanites searching for their roots submerged the original purpose of these events.

Enjoyment of these and so many other late-winter menu items in the early nineteenth century depended on people's acquired tastes. Ramps, for example, were a welcome visitor in the spring but left one with distinctively noxious breath for several days after consumption. The conservatism of our foodways is demonstrated in the continuation of winter and summer diets well past the widespread availability of low-cost preserved and imported foods. Heavier winter diets continued in millions of American homes even after World War II. My childhood home winter menu included seemingly endless Mason jars of stewed tomatoes that had been canned in summer. I still get queasy when I see a serving of stewed tomatoes heading toward my plate. My brother, in contrast, remembers this winter dish quite fondly and consumes it with relish as a comfort food.

Summer was a time of feast; rural families consumed increasing amounts of fresh foods, though much of what we consider the heart of the summer diet had not yet been "invented" for the nineteenth-century American kitchen. Lettuce and salads were served, but the lettuce was generally "wilted" (lightly sautéed) and served with hot dressings usually based on bacon drippings. Lettuce salads garnished with tomatoes, cucumbers, carrots, and the like are largely of twentieth-century vintage. Broccoli, asparagus, artichokes, and numerous other vegetables were known but little consumed until after World War II.

The gap between urban and rural diets began early and has continued through contemporary times. Although rural Americans typically ate far better than their urban counterparts, urban dwellers had a far more varied diet. This was especially important in the winter, when food could be imported from warmer climes. Both diets, however, were quite monotonous by modern standards. All but the very wealthy suffered from dietary-deficiency diseases, including malnutrition. Part of this stemmed from the fact that vitamins and their role in nutrition had not yet been discovered, though the abject poverty and meager diets of large numbers of urban residents certainly played an even greater role for them.

The Structure of American Meals

Most Americans attempted to eat three meals a day during the colonial era: breakfast, dinner, and supper. Breakfast generally was a light meal, often consisting either of a grain porridge or leftovers from the previous day. Dinner, the midday meal, was traditionally the most important meal of the day in medieval Europe and continued as such in America. This meal generally offered those at the table the most variety. Supper was a light meal and consisted primarily of food left over from the dinner table. A late afternoon "tea" started becoming common during the Renaissance among some economic classes in Britain but was comparatively rare in America except among those trying to emulate the British.

The evolution of American meals was largely shaped by the technology of the time. The lengthy process of getting a large log burning properly in the fireplace, heating water, preparing foods for the pot, waiting for slow-cooked foods to be done, and the myriad other duties that needed to be completed in order to prepare a large meal encouraged most Americans to concentrate on cooking one large meal a day and confine the others to either simple foods, such as porridge or pancakes, or leftovers. Extensive breakfasts of several meats and an assortment of other dishes were consumed by those who had a staff to prepare them. The remainder of Americans had to wait for the invention of the wood-fired iron stove to expand their breakfasts to include more dishes.

This pattern of cooking large quantities for dinner with the intent of consuming leftovers at supper and possibly even breakfast is reflected in the recipes of the day. Quantities tended to be so enormous that no family was expected to consume the entire dish in a single meal. The chicken pot pie recipes included in Amelia Simmons's (1796) and Eliza Smith's (1758) cookbooks, for example, both started with "pick and clean six chickens." Much of the meat of the times was boiled, typically taking several hours; beef hung over an open fire took hours to be properly cooked. Finally, cooking in an open fireplace was a challenge. The Dutch oven or pot had to be moved as the fire changed; a piece of beef had to be constantly turned and positioned at the proper distance from the flame. A bit of bacon was sometimes fried for breakfast, but this treat was more likely prepared when a family had guests and visitors.

The transformation of dinner into a lighter meal began in the early nineteenth century for urban residents. The development of factories employing hundreds of people at a site far from home meant that thousands of Americans were forced to alter their traditional midday meal. Some carried a light meal with them; others purchased food from street vendors; and a few had the financial resources to purchase meals at taverns, boarding houses, and the restaurants that were beginning to appear in cities. Whatever the solution, the midday dinner was transformed for the working class for all time, and workers began to eat a more substantial and earlier evening supper. Factory workers had little patience to wait for a fashionable late evening supper when they had been unable to have an adequate midday meal.

The term "lunch" began to replace "dinner" as the name for the midday meal in America during the early Industrial Revolution. "Luncheon" is derived from a Spanish word that referred to a small piece of food, though it had come to refer to a light meal by the seventeenth century. The term was first shortened to "lunch" in print in 1812; lunchrooms, restaurants serving light midday meals to urban workers, began appearing in American cities in the 1830s. The term "lunch counter" first appeared in print in 1869. The transformation of the midday dinner to lunch took place much more slowly on the farm, where the heavy midday meal continued well into the twentieth century. Its replacement often coincided with the lengthening of the school year and the consolidation of school districts, which increased both the time the children were in school and the distance they were from home.

Finally, it should be noted that most colonial Americans did not come equipped with dinner service for twelve. Most families had only two or three pots for cooking and few more bowls and plates than were needed to feed the family at a given meal. John Winthrop, for example, told his wife to pack "2 to 3 skillets of several sizes, a large frying pan, a small stewing pan, and a pan to boil pudding" for the trip to Massachusetts (Winthrop, 1996). This suggests that the meals were simple and often served from the pot. Utensils were not much more common, and it was not until metal utensils were mass produced that separate serving utensils and full sets of knives and forks began to appear on estate inventories of any but the wealthy. Although seemingly a bit of trivia about the American way of eating, this lack of "proper" utensils also had an effect on what was cooked and how it was served.

The Evolution of American Regional Cuisines

The American diet began taking on a distinctive character almost from the beginning as New World crops were integrated into colonial diets. Traditional European recipes and preferences were modified permanently because of the new foods that were available; corn, turkey, pumpkin, and other New World crops also quickly began appearing on European dinner tables. A comparison of eighteenth-century American and British cookbooks suggests that there was much less change in New England than in either the Midland or the South. Some specific attempts were made to "Americanize" the diet after 1770, most notably the substitution of coffee for tea, but much of this apparent change was more in name than recipe. One of the most profound changes was the replacement of rum by whiskey, but that was more related to economics than politics.

Growing regional differences in diet were even less self-conscious than the overall Americanization of the nation's food preferences. There apparently was little overt desire to create something distinctively southern or of New England. The growing recognition and pride in these regional differences did not become well articulated until many of the regional preference patterns were actually in decline. It is not suggested that southerners and residents of other regions did not recognize that their diet was distinct from the remainder of the nation at the turn of the twentieth century; rather, it is believed that they made no attempt to gain a sense of geographic identity by consuming grits or baked beans. In fact, it would appear that many of the nation's elite did not accept these rude American foods and ways of dining at all but continued consuming a largely European diet well into the twentieth century.

New England

The South and rural New England had the most British of menus, as their populations were largely of British origin; Pennsylvania and nearby areas tended to have populations with more varied ethnic origins and more distinct diets. The vast majority of sixteenth- and seventeenth-century immi-

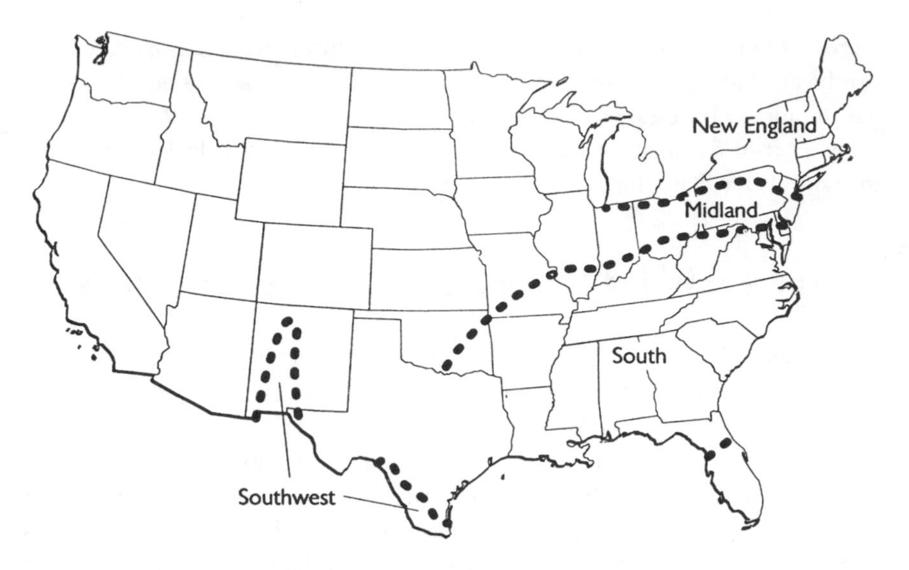

Traditional American Diet Regions

grants to New England came from eastern England, and their regional cuisine reflected the homogeneity of their heritage by retaining its strongly British flavor longer than any other regional cuisine in the colonies. New England cuisine was far from uniform, however; subregional differences existed there as they did elsewhere.

Proximity to the sea was the largest single factor influencing diet in New England. Amelia Simmons devoted four pages of her cookbook to telling homemakers how to select fresh fish and describing some of the tricks utilized by unscrupulous fish dealers. Interestingly, she did not include a single recipe for the preparation of fresh fish, though there were directions on how to prepare salt cod. This may have been oversight or may simply reflect her inland location in Hartford. Despite this oversight by Simmons, it appears that fish were one of the single most important elements of the diets of those living near the coast. Whether fried, boiled, cooked "alamode" (stewed), or chopped into pies and fish cakes, fish was included in nearly every description of a large dinner in a coastal area. A good part of the fish consumed both in the interior and along the coast, however, continued to be salt cod through the middle of the nineteenth century.

New Englanders ate more beef than other regions during the colonial period, though salt pork continued to be the most important meat product among the lower and lower-middle classes. An analysis of seventeenth- and eighteenth-century estate inventories indicated that salt pork dominated the food inventory in the seventeenth century and that the amount of cured beef consumed began increasing in the eighteenth century. Estate records did indicate that improving economic conditions did increase meat consumption, as reflected in the meat allowances provided for widows, which increased from about 80 pounds annually in the late seventeenth century to over 200 pounds in the early nineteenth century (McMahon, 1985). Cold New England winters meant that more meat could be kept fresh (or actually frozen), and farmers would have been more likely to slaughter larger animals during this time because of the lack of sufficient winter fodder. New Englanders were also the largest consumers of mutton, especially after the development of the woolens industry, which increased the numbers of sheep grazed on otherwise unusable land.

New England families never really accepted cornmeal with enthusiasm. But small fields and a short growing season combined with a wheat blight after the middle of the eighteenth century reduced wheat production, and cornmeal consumption greatly increased despite their preferences. Rye and oat production increased in the late eighteenth century, but cornmeal consumption remained at comparatively high levels for most interior residents until the development of cheap overland transportation in the nineteenth century. A colonial account book from Deerfield, Connecticut, suggested that wheat flour was consumed over cornmeal at a rate of about four to one in the mid-eighteenth century. Rye flour was substituted for wheat in areas where it was an important crop. Boston brown bread, which was made with a combination of wheat or rye flour and cornmeal, continued to be popular throughout the nineteenth century because it was tasty and required only a small quantity of expensive wheat flour.

Turnips, onions, cabbage, carrots, parsnips, and a host of beans, pulses, and legumes were all grown from English stock in New England gardens and played important roles in colonial daily meals. To these were added New World crops, such as sweet corn and pumpkins, to create a comparatively broad-based group of foodstuffs. Baked beans and succotash may be the closest to signature dishes for this region—one based on Old World traditions and the other on those of the New World. Succotash, an Indian dish of corn and lima beans, actually was found throughout British colonial America, though it is most often associated with New England.

Colonial New Englanders were heavy drinkers who consumed large quantities of beer and rum from the outset. Interior Connecticut was one of the hotbeds of sedition during the Whiskey Rebellion, and whiskey was commonly consumed in the interior after 1750. Hard (fermented) apple cider, a

Boston Brown Bread

1 c. rye meal	*1 c. corn meal*
1 c. coarse entire wheat flour	*3/4 tsp. soda*
1 tsp. salt	*3/4 c. molasses*
2 c. sour milk	

Combine and sift dry ingredients. Add molasses and milk and stir until well mixed. Place in well-greased mold not more than 2/3 full. Cover closely and place in kettle with boiling water coming halfway up around mold. Cover closely and steam 3 1/2 hours. Take from kettle and place in slow oven for about 15 minutes to dry off.
(Based on several recipes about 1910)

favorite beverage in many parts of England, increased in popularity as apple production expanded in the eighteenth century. Inventories, store accounts, and other sources also make frequent references to imported wines, brandies, and other grape-based drinks from the seventeenth century onward, especially among the more affluent. It was only the onset of prohibitionism in the late nineteenth century that this heavy consumption began to wane.

The New England breakfast typically was little more than porridge, cornmeal mush, or leftovers from the previous day. Numerous accounts exist of diners purposely saving favored parts of their supper for breakfast. Dinner thus not only was the largest meal of the day but set the tone for the next meal or two, and virtually all of the recipes of the time reflected this intent.

Far more meat was consumed in New England than in Europe during this period, but the medieval potages, most notably represented by Boston baked beans and succotash, played a continuing and important role in the daily diet, especially during the first century or so, when food supplies often were limited for most of the lower classes. Numerous recipes for fish cakes and hash exist, but these dishes typically consisted of cold boiled fish fried with chopped or grated potato, spices, and a little salt pork for fat and flavor. Suppers tended to be a lighter replication of dinner.

Midland

Midland cuisine was the most varied in origin of any of the traditional colonial regional cuisines. The earliest European settlers in the Hudson and Delaware River valleys were from the Netherlands and Sweden. The English quickly gained political control of these areas but did not dislodge the earlier Dutch settlements, which continued their lifestyles and landscapes well into the twentieth century. The remainder of upstate New York and a bit of northern Pennsylvania were settled primarily by westward-moving farmers from New England and, from a dietary standpoint, represent a prong of later New England settlement. Long Island and East Jersey were settled initially by southward-moving New Englanders.

The West Jersey colony (actually in southern New Jersey), in contrast, was primarily settled directly from England by Quakers in search of religious freedom. William Penn, one of several trustees of the West Jersey colony, became disturbed that only Quakers were welcomed in that colony and finagled ownership of Pennsylvania in 1681. He made this America's first colony that was open to anyone who could get there. The result was the development of a Midland cultural community that was somewhat in variance with the traditional historical view of the Middle Atlantic. The Midland, sometimes called the Pennsylvanian region, consisted of most of Pennsylvania (excepting a northern tier of counties settled by New Englanders), southern New Jersey (the West Jersey colony), and much of Maryland (excluding the Chesapeake Bay area). Large parts of western Virginia were primarily settled by emigrants from these areas, and as a result, much of the Shenandoah Valley and areas to the west have strong Midland characteristics. Word preferences, standardized house forms, dietary preferences, and a host of other cultural elements remain as silent reminders of this early cultural mixing along the Virginia frontier and in eastern Ohio.

William Penn's liberal Quaker attitude attracted a wide variety of immigrants, most notably large numbers from central Europe, nominally called Germans or Pennsylvania Dutch, and Scotch-Irish (Protestant Irish who had previously moved to northern Ireland from lowland Scotland). Most of the German settlers were religious dissidents, primarily Lutherans, Brethren, and various evangelical groups. The highly visible Amish and Mennonite sects were never more than minor elements of the large Germanic population in this region.

The first group of Germans came to Philadelphia in the last decade of the seventeenth century and settled at Germantown in what is now the suburbs of that city. Southeastern Pennsylvania became the largest center of German settlement in America at this time. About 38 percent of the population of Pennsylvania was of Germanic descent in 1790; less than 10 percent of the nation's total population was of that heritage (Purvis, 1987). Studies of colonial Pennsylvania ethnicity suggest that most Pennsylvania Dutch lived in distinct ethnic districts, primarily in the counties along the Maryland border or just to the north. More mixing did occur as the southeastern Pennsylvania culture expanded westward into the ridge-and-valley region of central Pennsylvania and the Cumberland and Shenandoah Valleys of Maryland and Virginia, but even here there was far less intermingling of ethnic groups than might be expected after two or three generations.

This concept of minimal mixing can easily be overemphasized. Intermarriage and interaction between the major ethnic groups in the Midland were far more common than is generally assumed. An interesting illustration of the way that a culture amalgamates cultural elements without awareness is seen in a 1938 interview of an apparently British-origin respondent: "There was one custom Grandma never failed to keep, and all of her daughters kept it up, and one of my sisters still does to this day—that was to fry 'fox-knots' every Shrove Tuesday, because if you didn't fry them on that day and use plenty of lard in the pan, then you would not get enough lard when you butchered to last you through the year" (Perdue, 1992). The classic Pennsylvania Dutch custom of preparing fasnachts on Shrove Tuesday morning to ensure a bountiful harvest had been continued without any recognition of its origins. Although it is possible there may have been some Pennsylvania Dutch heritage in this family's genealogy, it is more likely that the custom just passed into the family without any awareness of its origin or the transliteration of the terminology.

In the eighteenth century in southeastern Pennsylvania, a Pennsylvania Dutch cuisine evolved that was primarily of central European origin with little admixture of American ingredients. The Pennsylvania Dutch never took to corn as a bread grain, though cornmeal mush and corn pancakes were regular fare. The Pennsylvania Dutch and their cuisine have become so glorified that southeastern Pennsylvania, especially Lancaster County, has become virtually unrecognizable as foraging urban weekenders comb the area for cultural experiences.

Elizabeth Ellicott Lea's cookbook well illustrates that a largely common dietary regime evolved in this area despite the apparent continuation of ethnic separation (Weaver, 1982). Lea was a staunch Quaker homemaker from Sandy Spring, Maryland, and wrote a cookbook that had a decidedly British cast; yet there were large numbers of Pennsylvania Dutch recipes, including "cold" slaw, bologna sausage, bacon dumplings, scrapple, apple butter, and pickled cabbage. Many standard British recipes had also acquired a hint of the Germanic influence.

In the region's colonial and early nationalist period, diets were heavy by any standard. The afflictions associated with high cholesterol must have been a major problem among those who survived to old age. The Germans fried too much of their food. It was often greasy and always salty. Salt pork was the preferred meat. Butter and lard were used extensively in many dishes.

The German tradition of seven sweets and seven sours on the table was especially conducive to heavy meals. Not every table had all of these at every meal, but several desserts were available at every meal among those wealthy enough to support this dietary lifestyle. Finally, the Germanic tradition of the "groaning board," that is, serving family and guests more food than they could possibly consume, also contributed to the heaviness of these meals and those who consumed them.

Breakfast tended to be the lightest meal of the day, generally focusing on fried or plain mush. Oatmeal porridge was also popular among both the Pennsylvania Dutch and the Scotch-Irish populations. Other breakfast favorites were waffles, buckwheat pancakes, scrapple (less often fried bacon, sausage, or salt pork), and various breads. The potato had been introduced into German cuisine before the Pennsylvania Dutch came to America. Fried fresh potatoes, not hash browns, was a favorite dinner treat. Journey cake and an assortment of other corn dishes were known to the Pennsylvania Dutch but were more often consumed by their British neighbors. Milk toast and chipped beef and gravy on toast were also less common among the Germanic settlers. Guests ate even better, making travelers' accounts of meals in the region extravagant even by local standards.

Dinner was the largest meal with a host of dumpling, noodle, and potato dishes appearing on the midday dining room table. Chicken pot pie, a standard (though not necessarily frequent) entrée of the entire East Coast, was very popular, though the recipes varied widely even over short distances. Salt pork was the single most important source of protein in the preindustrial era. Families were large and usually multigenerational; the grandfather and grandmother often lived with one of the children after retirement. Large numbers of children and hired or volunteer hands rounded out the large group that sat down at the dinner table. Several entrées usually graced the table as well as an assortment of sweet and sour relishes, one or more breads, and side dishes. Onions of all kinds made their way to the table, especially in the winter, as did sauerkraut among the Germanic families. The British families consumed sauerkraut occasionally but more frequently boiled their winter cabbage or consumed it in a slaw.

Supper was a lighter meal than dinner and, as in New England, was typically created from the leftovers from earlier in the day. The region is widely known for its sausages, and these were a favorite at the evening repast if the leftovers were supplemented by newly cooked dishes. Lebanon bologna,

Chicken Pot Pie

The general consistency of some recipes reflects both the borrowing among cookbooks and the simplicity that characterized most recipes before the "big" cookbooks of the Boston Cooking School and other sources. It seems that chicken pie was chicken pie wherever you encountered it in nineteenth-century America; it was a bit rich and greasy by today's standards.

Chicken Pie (New England, 1796)

Pick and clean six chickens. Take out the innards, wash the birds, joint them, and salt and pepper both innards and pieces. Roll one inch thick paste No. 8 and cover a deep dish. Put thereto a layer of chickens and a layer of thin slices of butter until the chickens and one and a half pounds of butter are expended. Cover with a thick paste. Bake one hour and a half.

Paste No. 8: Rub in one and half pounds of suet to six pounds of flour and a spoon full of salt. Wet with cream and roll in two and a half pounds of butter.
(Simmons, 1796)

Chicken Pudding, a Favorite Virginia Dish (1825)

Beat 10 eggs very light, add a quart of rich milk with a pound of melted butter and salt and pepper. Add flour to make a good batter. Take four young chickens and clean. Cut off legs and wings. Boil in a sauce pan with thyme and parsley until nearly done. Place chicken and batter in a deep dish and bake. Send nice gravy in a boat.
(Randolph, 1825)

Chicken Pie (northern Maryland, 1853)

Cut up the chickens and boil them fifteen minutes in a little water if they are old. Make a paste of common pie crust and pat it round your pan or dish. Lay in the chicken, dust with flour and put in butter, pepper, and salt. Cover them with water. Roll out top crust quite thick and close around the edges. Make an opening in the middle with a knife. Bake an hour. Pour off the gravy and warm it separately if you reheat the next day.
(Weaver, 1982)

almost entirely considered a sandwich treat today, was fried and served as a meat, as were smoked and other sausages, especially scrapple. A few fried potatoes, a little bread, and maybe some dessert and supper was ready.

The South

The "South" has always been as much a state of mind as it has been a place. Simplistic attempts to find the outer limits of this place too soon bog down in conflicting and contradictory evidence. Certainly, few characteristics are endemic to the entire region, and the boundaries of the eleven states of the Confederacy were never culturally relevant. The South defies easy delimitation because it has never had a central core area spawning new concepts and traditions and sending out pulses of support to keep them regionally pure, as did Philadelphia and Boston for their respective dependent regions. Indeed, southerners have always been self-absorbed and oriented toward that narrow home place in which their lives are invested. There were no large cities other than some small coastal entrepôts until the rise of Atlanta; nor have their ever been any individual voices that were able to speak for the entire region. The result has been the development of a highly compartmentalized set of interlocking subregions, all linked to a larger way of life yet each quite distinctive in its own right. The complexity and disjointedness of this place is its central theme.

Traditionally the South has been divided into the Upland and Lowland regions. The Upland South, roughly composed of the Appalachian states and flanking areas, was settled by a combination of southward-moving Midlanders with admixtures of poor, westward-moving, piedmont southerners. Most farmers in this region find conditions harsh; a few pockets of wealth exist in the Kentucky bluegrass area, the Nashville basin, and a few larger river valleys. The farming, the farmers, and the region are poor. The majority have always been poor and today continue to remain largely isolated from the mainstream of American life. Independent in religion, occupation, and lifestyle, these people have resisted all change.

My first visit to the Upland South was with a fellow student at Louisiana State who took me home to visit his family near Williamsburg, Kentucky. Dropping in unexpectedly on his father at dinnertime in nearby Emlyn, we were treated to a supper of pork chops, creamed corn, corn muffins, and leather britches. The family had left isolated Goose Lick about twenty years before to find work, and it became obvious that the fare in this household had changed little from that described by Sam Hilliard in his study of antebellum South foodways. In the following week we never ate out, never had factory-processed food, never even saw beef served as a meat dish, listened to a lot of gospel music, had more than our share of moonshine, and were made to feel welcome in every home we entered whether it had running water, indoor plumbing, or enough of the food offered us if we arrived at mealtime.

The Lowland South is far from homogeneous and is most conveniently broken into three major units and two smaller ones. The Atlantic and Gulf coastal plains are much alike, though typically the later-settled Gulf coastal area had larger slave populations and much stronger African influences in its diet. (Central and southern Florida were largely unsettled by Europeans until the late nineteenth century and are not a part of this discussion.) The two subareas, the Sea Island coast of the Carolinas and Georgia and the Louisiana portion of the Mississippi delta are quite distinct and interesting dietary subregions; understanding their peculiarities helps us understand the dietary preferences of the region as a whole.

There probably has never been a truly "southern" diet, though there are some distinct preference patterns that distinguish this region from the cuisines outside the South. The poor had the most meager and least varied diet of the region (Hilliard, 1972). It is generally accepted that the African American population and the region's poor whites consumed much the same diet during most of the nineteenth century. Various forms of corn products provided the bulk of their caloric intake. Cornmeal was made into an amazing variety of breads and used as a thickening agent in many "stews." Generally about a peck of cornmeal per week per person was given to the plantation slaves (except on the "rice coast," where the staff of life was often rice). Those numbers obviously varied widely.

The poor consumed comparatively small amounts of meat and most of that as salt or smoked pork. Slaves, freemen, and poor whites alike most frequently consumed fatback, bacon, and the other less meaty parts of the pig; the elite tended to consume the majority of the hams and roasts. Chickens were common in virtually all farmyards, including those of the slaves, but as elsewhere likely were consumed primarily on special occasions. Eggs were an important nutritional supplement, and the money that could be obtained by selling a chicken that had stopped laying was greater than the value of the meat to a poor family.

Low rural population densities and the development of most of the larger urban areas near bodies of water meant that game continued to be a much more important protein supplement for these people than elsewhere in the nation. The shores teamed with shellfish, shrimp, and finned fish; catching catfish, bream, buffalo, and other freshwater fish continues to be a common after-work and Sunday activity for the poor. Large numbers of squirrels, possums, rabbits, and other small animals thrived in the lightly settled woodlands of the South and were a common meat supplement for many rural poor.

The vegetative diet evolved from the complex cultural history of the region. European crops were quickly supplemented by both New World and African foods. Corn became the core of the diet, consumed not only as cornmeal but also as an additive to a variety of stews, as fresh corn on the cob, and of course as hominy. A variety of New World crops were carried to Africa and seemingly brought to the United States from there. Intense debate has raged over their introduction to the United States, whether they were brought directly from Africa or brought by slaves imported from the Caribbean. This debate is largely irrelevant here, as there was so much traffic from both Africa and the Caribbean that all of these foods were probably introduced from both of these places at some time and in some places.

The Sea Island coast of the Carolinas was distinguished from the remainder of the South by both its climate and its demographic history. Though it is at the same latitude as most of the remainder of the region, the presence of the Gulf Stream just offshore moderates winter temperatures, allowing the cultivation of many subtropical crops including rice, indigo, and tea. This salubrious environment and open society fostered the immigration of large numbers of French Huguenots to Charleston and the surrounding region in the eighteenth century. It has been estimated that as many as 45 percent of the entire European population in coastal South Carolina were Huguenots, primarily from the Provence section of France. Some, if not many, are thought to have been Jewish refugees from Spain and Portugal. Large numbers of planters also had resided in the Caribbean prior to coming to the rice coast. Whether British, French, or from some other area, these Europeans brought a much broader based agricultural and culinary history than was found in other British colonies.

A large portion of the Africans brought to the Sea Island coast also had resided in the Caribbean prior to their arrival in America. The labor demands of the rice plantations were so great that Africans outnumbered the Europeans outside of the larger towns, reaching a ratio of nine to one or even higher in the lower Waccamaw (South Carolina) River delta and similar salt marsh areas of South Carolina and Georgia. The rice plantations tended to be operated on the "task" system. Groups of workers were given tasks that often required long periods of intense labor, but they were allowed a degree of freedom to farm their own plots, fish, or otherwise find uses for their own time between task assignments. Coupled with the practice of allocating comparatively large areas for slaves to cultivate their own food, these high densities of Africans led to a much greater influx of African food preferences into the local diet than on the tobacco and cotton plantations, where gang labor was typically used. Many specialty foods—especially sweet potatoes, sorghum, okra, peas, and peanuts—that would not have been available otherwise were grown and became important parts of their menu regime. It also was not uncommon for them to farm a bit of rice for their own use.

Africans often served as cooks and kitchen help on many plantations and in middle-class homes of the rice region. These cooks apparently received somewhat more freedom in menu development than African cooks in other areas, and a variety of basically African dishes entered the Carolina cuisine—often apparently without knowledge of their origins within the accepting European population. The signature dish of the Carolina rice country is pilau, a rice dish served with chicken, shrimp, or other meat. Pilau is probably of Persian origin; many writers believe that it was introduced into the Carolinas by French Huguenot settlers from Provence. Although this may be true, Karen Hess made a powerful argument in her analysis of the Carolina rice kitchen for West Africa as the origin of the dish (Hess, 1992). She pointed out that West African farmers had cultivated yams, peanuts, sorghum, okra, and, most important, rice long before their transport to America. They utilized a double-stage process in rice preparation that leaves the rice grains swollen but each distinctly individual. Proper Carolina pilau is made with rice cooked in this manner and almost always with a bit of bacon in the pot for flavoring—a most unlikely garnish for either Persian Muslims or Jewish Huguenots but one quite consistent with West African and West Indies cooking.

The Africans along the rice coast ate rice often during the slave period and at almost every meal afterward. They created hundreds of variations of their basic rice recipes. A common dish in West Africa utilized chickpeas and rice and became the second classic dish of the region—hoppin' John. The origin of the name is lost, but the dish and its hundreds of variations are known throughout the Lowland South. The Louisiana version, known as red beans and rice, has become almost an icon of the southern marshlands. A host of other African dishes, including gumbo and jambalaya, were added through time to the classic Charleston cuisine, suggesting that both of these dishes did not originate in America but in either the West Indies or West Africa.

It has been said that the three characteristics of southern cooking have traditionally been the frying pan, grease, and overcooked food. Like most gross generalizations, there is some truth and some hyperbole in this statement. Although frying is common, pots of cooked vegetables, stews, and vegetables are actually the hallmarks of this cuisine. Indeed, much southern cooking is

Hoppin' John

2 c. raw cow peas	*1/4 lb. jowl or fatback*
1 sm. onion, chopped	*4 c. water*
2 tsp. salt	*2 c. raw rice*

Soak the peas overnight in 3 c. water. In the morning, add more water and cut up jowl. Boil peas in salt water and meat about 30 minutes or until just tender. Do NOT cook to a mash. Add the 2 c. of peas, onion, and 2 1/4 c. of pea liquid with meat to 2 c. of rice. Put in steamer and cook 1 hour or until rice is done. (Harrigan, 1983)

greasy, primarily because the traditional southern cook tossed a dollop of fatback or streak o' lean into the pot to improve the flavor. Actually the only foods that are typically overcooked by national standards are vegetables prepared in large quantities and consumed over several meal periods. Because it was impossible to protect the food from bacterial growth, southern cooks continued the medieval European tradition of keeping the vegetable pot warm on the hearth or stove until it was empty.

The southern breakfast in its purest form is composed of fresh cornbread (as sticks, muffins, or johnnycake) with a bit of sorghum or cane syrup and perhaps dipped buttermilk. Alternatively, one could be served grits, a hot cereal made from ground hominy, or one could eat cornbread and grits. The most meager of breakfasts would be a hot yam taken from the ashes of the farmhouse fireplace. Millions of poor southerners, black and white, have had this as their only breakfast over the years.

Hundreds of other items could garnish this base meal, especially with expanding economic resources. Streak o' lean, sausage, or southern-cured ham were often fried and served in later times. Biscuits became a southern staple as wheat flour came into economic reach of the majority of the population, probably in the early twentieth century; today, most assume that they were always a tradition among all southerners. Eggs, of course, are an American breakfast tradition, as are assorted jams, jellies, and local delicacies.

Again, the midday meal was the largest meal of the day and typically included a variety of cooked vegetables and meats. About five pounds of pork were consumed for every pound of beef in the South with the poor consuming little or no beef at all. Most dishes thus were based on pork or vegetables. Peas (crowder, black-eyed, etc., but rarely English green) were served at most meals, as well as cornbread and, later, biscuits. Most families had only a single pot or two to use in the fireplace; hence modern ideas of the variety available then are largely inflated. Sorghum or cane syrup, often locally made, was also almost always available. Buttermilk too was frequently served, especially to children, as well as some sort of a coffee brew.

Supper typically was a lighter meal. If leftovers were not available, it was back to basics—cornbread, buttermilk, a bit of fried salt meat. Yams were always easy to prepare and available to the men coming back from the fields and the children around the house. There were many periods of hard times in the South, and hunger and malnutrition were common for long periods. Supper was often a meager meal.

Upland southerners tended to have more Midland elements in their diets, reflecting the heritage of many who lived there. The southward-moving Scotch-Irish, almost all of whom entered the region through Pennsylvania, were the most notable of these influences, although English- and Germanic-surname families also moved into the Upland in great numbers. The biggest difference in their diets was the lesser dependence on foods with African overtones. Yams and peanuts were rarely consumed by Upland families. Cool-weather vegetables—cabbage, string beans, white potatoes, and the like—were more common; collards, black-eyed and crowder peas, all of African origins, were less common. Similarly, apples were of greater importance in the Upland and uncommon on the coastal plains.

The French triangle of southern Louisiana has one of the most distinctive subregional cuisines in the nation. Famous for its settlement by displaced French colonials from Nova Scotia and the Caribbean, the area's cuisine was also influenced by immigrants from around the world because of the role of New Orleans as the gateway to the North American heartland. Typically this area is portrayed either as a bustling seaport or as a region of poor Cajun farmers, fishers, and trappers living along the bayous. Those images are only part of the complex cultural mix near the mouth of the Mississippi. There were also great sugar plantations throughout most of the region's history, rice farming in the late nineteenth century, and a resident Amerindian population. The cuisine was never pure French; nor were the people. New Orleans initially attracted people from throughout the Caribbean; later, empty cotton ships brought immigrants from many lands at bargain prices. Many of these people moved on to the interior when they could, but enough stayed in New Orleans to make it as cosmopolitan as any city in the nation. The plantations also required large numbers of slaves. Most were of West African heritage but often were brought here from the Caribbean or elsewhere in the South.

The region's cuisine as a result is a mixture of primarily French influences with hundreds of other elements from around the world, including some from the nearby traditional American South. The signature dishes of the region are its various fish and vegetable "stews," including jambalayas, gumbos, and étoufées. The signature ingredients are oysters, crawfish, shrimp, beans, okra, tomatoes, and rice. The combinations are endless.

The origins of jambalaya are controversial, although a strong argument has been made that it is a Louisiana adaptation of the same African dish that served as the basis for the hoppin' John so popular in the Carolina rice country. Considering the strong ties of both regions to the French Caribbean, it is possible that the word "jambalaya" entered the Louisiana patois from either or both sources. The oldest recorded use of the term was not until the late nineteenth century, about the same time that rice was being introduced into southern Louisiana from the Carolinas.

"Gumbo" too is an African term; its identifying ingredient, okra, also was introduced to and popularized in the American South by Africans. Gumbos typically are more liquid than jambalaya and contain a greater variety of vegetables but always okra, their signature ingredient. Like jambalaya recipes, those for gumbo vary significantly through time and in different areas. Étoufée, which was not originally found in the Carolinas, is quite similar to both gumbo and jambalaya, but its dark rue base is usually used as a gravy for crawfish or other meat or fish. All of these dishes are served with a side of white rice, the grains each standing perfectly alone in the African–West Indian tradition.

Virtually all Cajun cooking contains some form of game, especially fish, and is highly spiced with local peppers. An assortment of French-style pastries and baked goods are also found in the region, as well as the French version of coffee. My personal introduction to Louisiana coffee took place in a Baton Rouge rooming house when my new neighbor from Lake Charles offered me a mug of coffee and then fretted the entire time we drank it because he could still see the spoon blade when it was placed in the mug. I thought he was joking.

The Southwest

The pre-European residents of the New World had the greatest impact on the cuisine of the Southwest. The harshness of the arid environment limited local production of traditional European humid-climate crops. The great distances from sources of European foodstuffs and the frequent role of Hispanos* as cooks aided in the infusion of a wide variety of local foods into the pre–Industrial Revolution diet of the American Southwest.

The Southwest's pre-European diet centered on a duo of corn and beans with a variety of other vegetative ingredients, most notably chilis and squashes. Corn was most often ground into meal and was made into flat tortillas, which served as a very versatile bread product. Even after wheat flour was introduced to the region in the nineteenth century, the tortilla continued to be formed and cooked in the same manner as before. Beans were cooked in a variety of ways, often with other vegetables, and thus the most traditional pre-European meals consisted of a stewlike dish of vegetables, possibly including meat, served with tortillas. Thick, almost milkshake-like concoctions of water, corn, and flavoring ingredients were a very common breakfast among the Maya and were well known northward among the inhabitants of this region as well. Enchiladas and tamales also have pre-Columbian origins. All of these dishes have regional characteristics along the Hispano borderland. The Texas tamale, for example, is about the size of a man's thumb; those in Arizona and California typically are about the size of an ear of corn.

The transition to a hybrid diet after the arrival of the European ranchers was relatively simple. Beans became a central theme, especially among the working classes. They were served at least once a day, and it was not uncommon to have beans at every meal. Tortillas were not as popular among the new Anglo† residents, but cornbread was a favorite—many early Anglos being recent migrants from the American South. Local beef became an important part of the diet in this precholesterol cattle-ranching era among the Anglos; the recently introduced goat, which survives well with little care, has become common among the Hispano population. The Anglo and Hispano diets appear to have remained distinctively different throughout the nineteenth century, but actually there was acculturation and mixing almost from the first day of Anglo occupation.

---

*Hispano is used here to describe those people and their culture who had settled in the Southwest prior to the coming of the Americans. It is contended that these people are fundamentally different from the later Mexican-Americans who largely emigrated after 1965. Many Mexican Americans now live in some of the same areas as the Hispanos, making it increasingly difficult to distinguish the groups.

†The term Anglo is used in these discussions of the Southwest to denote any Euro-Americans regardless of their actual national origin.

Final Thoughts

The preindustrial American cuisine was complex and featured an almost endless array of social, economic, and regional patterns. Some clear patterns do emerge: (1) the diet remained largely European in character; (2) most Americans ate a very bland and monotonous repetition of dishes largely based on the same components day after day; and (3) a great deal of regional intermixing and amalgamation was taking place even before the days of the telegraph and the Internet.

Even a random examination of a few early recipes indicates that many dishes were much the same along the East Coast even though they often had local names. Chicken stews with corn or dumplings or rice were perennial favorites and tasted much the same despite the substitution of locally available ingredients. Great variations also existed even within some of the relatively small subareas, and these continued into the twentieth century. The transformation that was to take place over the next century would be massive; the standardization of the American cuisine was unprecedented.

The transformation of the American diet was of course paralleled by a transformation of the nation's culture generally. The new American cuisine of the late twentieth century was derived from these early beginnings, but it was the flood of immigrants from new areas and technological change that has defined it.

3

Stocking the Pantry: Technology and the Food Supply

> *My aunty . . . raise me up on his farm, was plant cotton, and peas, potatoes, and all. For dinner we eat peas and some corn grits, little sweet potato, and drink some water. . . . We see meat, but we don't get it.*
>
> **—Willie Hunter, Johns Island, South Carolina (Carawan and Carawan, 1989)**

The stark, repetitive diet of colonial America began to slowly change in the nineteenth century as growing urban markets became dependent upon food produced by others and as technological innovation worked its miracle on the American food system. Food became cheaper, and its quality improved remarkably in a few short years. This transformation should have just created more of the same at lower costs; instead it fostered a complete restructuring of the American diet.

Urban Growth and Commercial Food Production

Only about 200,000 Americans lived in towns and villages in 1790. There were only twelve communities with more than 5,000 souls; New York with 30,000 inhabitants was a small town by today's standards. Virtually the entirety of the remaining 3.8 million inhabitants of the new nation lived on farms, produced almost all of their own food, and had little disposable income for more than a meager supply of imported foods.

The urban system of the time was just too small to support the development of an economically efficient infrastructure for the movement of foods from region to region or the processing of large volumes of even basic locally produced foods to achieve economies of scale. Most towns were loosely settled, with many inhabitants cultivating home gardens while raising a few chickens and a milk cow or two. Money was in short supply, but most of those who did tend home gardens also purchased at least part of their food supply. True urban places, with the population increasingly forced into multiunit residences and little opportunity for at least some food production, did not begin to become important for another fifty years. Manufacturing employment remained small; fewer than 1 million factory workers were reported in the 1850 census, and the vast majority of those worked in water-powered mills in largely rural areas. Large-scale food manufacturing was dependent on the development of a large home market, and expansion in this area closely paralleled the urbanization curve; increasing numbers of people became totally dependent on purchased food supplies.

Technology and the Food Supply

Technological innovation permeated every aspect of nineteenth-century American life. The rapid staccato beat of change created an atmosphere where change was good and tradition was bad. Recent studies of agricultural innovation clearly demonstrated that American farmers integrated new technology into their agriculture well before the reaper, the tractor, and other innovations were more economical than their predecessors. New ways of farming, new ways of building homes and offices, new ways of transport, and new ways of cooking were all created and integrated into everyday life. The innovations creating the most dramatic changes in the American diet during this period included the development of the cooking range to allow the easy preparation of more complex meals; the creation of a low-cost, efficient transportation system; the creation of a variety of new food-preservation systems; and the early development of mass food-processing facilities and corporations to operate them (see Figure 3.1).

The Cooking Range

Cooking on the open hearth was one of the most disagreeable jobs facing the eighteenth-century cook. The blazing heat, popping embers, and the hour or two that it took to establish a proper bed of coals for cooking all made this a task to be avoided. The cooking stove, or range, as it began to be called after 1850, was thus an important innovation in making the task of cooking more convenient as well as allowing the home cook the opportunity to more easily prepare multipot meals. Of course, baking, preparing multicourse meals, heating water, and the remainder of the hearth chores had been accomplished for centuries prior to the introduction of the cookstove, but the efficiency, ease, and comfort of working at waist height instead of bent over pots in the hearth made all these jobs easier and more predictable in their outcome.

The history of the cookstove is shrouded in ambiguity. The first European stove, a brick and tile affair, is generally attributed to the Rhine River region of Alsace during the fourteenth century. A cast-iron version appeared about 1490 in the same area. Stoves also appeared quite early in Russia and Scandinavia, though it is not known when householders began developing versions

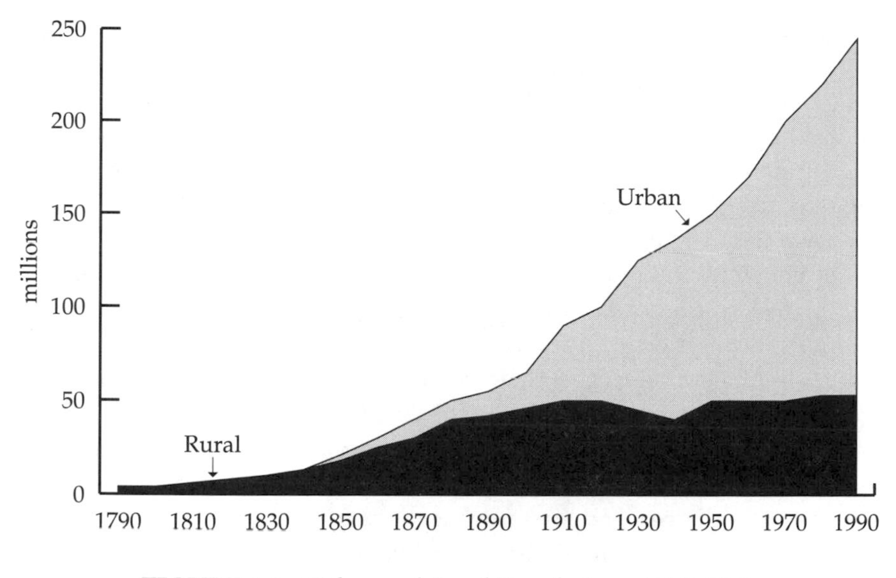

FIGURE 3.1 Urban and Rural Populations: 1790–1990

that would be used primarily for cooking. Heating stoves were widespread, if not very common, in Europe prior to the colonization of the United States, and most colonists would have at least heard of these new devices prior to their departure for the New World. *The Oxford English Dictionary*, for example, lists citations of stoves in England as early as 1562, with the pieces of a cast-iron stove listed for sale in 1618.

The first cast-iron heating stove manufactured in the United States was manufactured in Lynn, Massachusetts, in 1642. Cast-iron heating stoves were also imported from Europe by both the early Dutch and Swedish settlers along the Hudson and Delaware Rivers; there are many eighteenth-century accounts of Pennsylvania Dutch ironmasters creating five-plate heating stoves for placing in the backs of fireplaces to heat the room behind. Indeed, there has been some question whether the Pennsylvania Dutch were more predisposed toward adopting iron stoves than their British neighbors because of the widespread adoption of stoves on the continent prior to emigration. One study of eighteenth-century inventories found virtually no stoves in British-settled Burlington County, New Jersey, whereas 20 percent of the inventories of German-dominated York County, Pennsylvania, listed stoves. It is unlikely, however, that more than a handful of all of these stoves were used primarily as cooking devices. Further, an examination of eighteenth-century British home life indicates that stoves were better integrated into life there than in the United States. Two other conclusions may be drawn from the study of American estate inventories, however. First, frontier areas such as York County, Pennsylvania, where most homes were comparatively new, were more likely to have stoves than older, settled areas further east. Second, the diffusion of this comparatively expensive device was slow until the nineteenth century, when manufacturing economies of scale brought the heating stove into the easy economic reach of more and more families.

The adoption of the stove for household use was not without its dangers. In the 1841 edition of her widely circulated *Treatise on Domestic Economy*, Catherine Beecher devoted several paragraphs to stove problems. She noted that many found disagreeable the extreme dryness of the air associated with homes heated with stoves. Many also disliked the "disagreeable smell of the iron; and the coldness of the lower stratum of air, producing cold feet in those who are subject to that difficulty" (Beecher, 1841, 299). She also carefully brought the readers' attention to the ventilation problem when she noted, "Stoves for coal should be carefully put up, as, if the pipe gapes, the coal gas may occasion death, especially if it escapes into a sleeping room" (Beecher, 1841, 300). Despite these problems, on the whole the wood and coal cooking and heating stoves were perceived to be great innovations because they allowed for more convenient cooking and because some homes were actually warm in winter for the first time.

The introduction of the cooking stove into American life is poorly documented. Benjamin Franklin's invention of the Pennsylvania fireplace in 1740 brought attention to the use of iron stoves for heating. Isaac Orr of Philadelphia is often given credit for manufacturing the first cast-iron cooking stove in 1800. It used a set of grates over the fire rather than the later iron plates to allow direct heating of the pots. Oliver Evans introduced the oven and hot-water heater tank in 1806. These two additions immediately became mandatory elements of a proper cookstove, but the closed firebox did not begin appearing until the 1820s. A battle of words erupted between the advocates of open cooking grates and those favoring the newer closed cooking surfaces. Many believed that Orr's open-grate direct-fire heating provided faster and better cooking, partially because few had the flat-bottomed pots and pans needed to properly use the flat-surfaced, closed-firebox system. The enclosed firebox had won the battle by 1850.

Wood-smoke pollution and the provision of sufficient quantities of properly cured hardwood for cooking and heating were problems by the early eighteenth century in many urban areas. Coal did not burn properly in stoves manufactured for wood because of the inability to properly control the flow of air over the fire. Jordan Mott of New York invented a sliding-grate system that allowed the use of nut coal (small chunks of anthracite coal) in 1819. Unfortunately, most hard coal was expensive because it was imported from England, but his invention, coupled with the increasing use of anthracite coal in the previously charcoal-fueled iron industry, helped stimulate the development of the anthracite coal fields of northeastern Pennsylvania. Philadelphia investors quickly moved to create a canal system to the anthracite coal fields of eastern Pennsylvania to tap this new market, and nut coal soon become available at reasonable prices to most urban dwellers.

The Franklin Institute of Philadelphia began heavily promoting the use of nut coal for heating and cooking in the 1820s. Prizes were offered to the inventors of more efficient anthracite coal–burning stoves, and cost-saving successes were gleefully passed on to its readership. For example, the institute reported to its readers in 1825 that the Pennsylvania hospital of Philadelphia reduced its fuel costs by one-third when it switched from wood to coal for heating and cooking.

Little is known about the actual rate of adoption of cooking ranges by American households during the nineteenth century, and numbers and reports are conflicting. The 1850 census indicates that there were fifty-four foundries in thirteen states making stoves and ranges valued at slightly over $6 million. The 1860 census separated cooking ranges from heating stoves, noting that there were thirty-seven factories in four states employing 290 workers to make "hot air furnaces, cooking ranges, etc." Philadelphia was the largest production center with twenty-two foundries, followed by Massachusetts, New Jersey, and Ohio. New York was not listed as having any cooking-range factories but was the largest center of heating-stove production. The 1870 census indicated that 1,285,177 heating stoves, 15,351 hot-air furnaces, and 5,450 cooking ranges were manufactured in that year, suggesting that the bulk of production in 1850 and 1860 was not cooking stoves.

Large numbers of wood and coal cookstoves were also being manufactured in England and on the continent at this time, making it even more difficult to estimate the rate of acceptance and acquisition of cooking ranges during this period. Smith and Wellstood, one of Britain's larger manufacturers of cooking and heating stoves, was founded by James Smith, who spent nine years in Mississippi prior to returning to Scotland in 1841. Together with his childhood friend Stephen Wellstood, he formed the firm of Smith and Wellstood to make American-style stoves. In the 1880s the firm produced forty-three different types of heating stoves and cooking ranges; in 1912, the company manufactured more than 200 types. The American market for Smith and Wellstood, as well as other European stove builders, was an important element of their business.

The gas cooking stove began appearing in England in the early part of the century but did not come into wide usage in the United States until the end of the nineteenth century. There were twenty-four factories producing gas stoves in 1890, all located in the Northeast and Midwest.

Transportation Innovation

A variety of transportation-related innovations played a more obvious role in altering the food supply. The combination of rapidly declining transportation costs and the development of the means to haul ever larger quantities increased both the quantity and quality of foods available to all Americans. Imported wheat flour, rum, and other foods were available to coastal residents throughout the colonial period, as otherwise near-empty, inbound cargo ships kept freight rates low to help fill their vessels. Coastal residents could purchase a wide range of commodities from Europe—wheat, barley, wines, and white potatoes—and tropical and subtropical specialties from the Caribbean such as rum, molasses, sugar, and mangoes. The availability of these crops affected the economic viability of grain farming almost from the beginning.

These imported goods were also carried into the interior, especially if water routes were available. Records of a colonial store in Deerfield, Connecticut, for example, listed sales of rum, a variety of Portuguese wines, raisins, figs, lemons, assorted spices, tea, Jamaican and other sugars, and molasses. Rice, salt cod, a variety of other salt fish, salt, and hops were all produced outside the community, and some may have been imported. It is impossible to determine how much of the average person's daily food intake at this time was imported from outside each community, though Deerfield's annual per capita consumption of select imported foods was estimated to be 3.75 gallons of rum, 2.5 pounds of tea, and 5 to 10 pounds of sugar during the late colonial period.

The development of low-cost overland transportation in the early nineteenth century brought more imported goods to inland residents and made possible the shipment of produce to coastal cities as well. The completion of the Erie Canal in 1821 revolutionized the economics of production in the Midwest. A frenzy of canal building erupted. Freight rates dropped to as low as one cent per ton mile within a decade, further encouraging the expansion of western agriculture as well as forcing a restructuring of agriculture in New England.

Canal transport had problems, and the development of railroads soon challenged its supremacy even though rates were 1.5 to 2.5 cents per ton mile higher for rail transport. Canal boats were always slower, typically taking almost three weeks to move salt pork from Cincinnati to New York; the railroads took only about a week to move the same cargo between those cities. The *Niles Weekly Register* reported in 1831 that the Erie as well as the Delaware and Hudson canals had been shut down during that winter for five months due to freezing. The Baltimore and Ohio Railroad lost only a single day to bad weather during the same winter. Food shippers could not tolerate these delays, and soon they shipped little more than grain by canal boat.

In retrospect, the expansion of the rail system during the nineteenth century was astounding. Beginning with just 23 miles of track in 1830, the rail companies built almost 9,000 miles of track over the next twenty years and more than 20,000 additional miles in the following forty years. Rail freight tonnage expanded from 2.2 billion ton miles in 1865 to 17.8 billion ton miles in 1885. Simultaneously, freight rates declined precipitously, especially between high-volume locations, allowing the relatively inexpensive shipment of foods to almost every community in the nation. Thousands of freight forwarders, food wholesalers and distributors, and specialty distribution networks appeared almost overnight.

Natural Ice and Food Preservation

The ability to move foodstuffs long distances allowed for the shipment of grains and a handful of other crops that did not need preservation, but inexpensive transport had little effect on produce, meat, dairy, and other perishables. The expansion of these industries was dependent on the development of efficient and economical methods of retarding spoilage. Dried and salted meat and fish had been available for centuries, but it was the development of refrigeration, canning, and, later, freezing that opened opportunities for entrepreneurs to make large quantities of produce and other perishables available at reasonable prices.

Ice had been used for centuries for cooling foods, especially drinks, and it had been discovered by the end of the seventeenth century in France that a mixture of saltpeter and snow created very low temperatures that could be used to create iced liquors and frozen juices. Despite the wide use of ice for cooling drinks, creating desserts, and occasionally retarding spoilage of fish and other fresh foods, little was actually known about the decay process in food. Indeed, Francis Bacon apparently died of food poisoning as a direct result of his experiments on the preservative effects of snow stuffed into a dressed chicken. A few entrepreneurs did utilize the cooling effects of ice to ship fish on the Erie Canal and overland to urban markets such as Philadelphia and New York, but their numbers were relatively minor prior to the invention of machinery that made artificial ice.

Icehouses for the storage of natural ice became increasingly popular after 1800. George Washington had one constructed at Mt. Vernon in 1785, and Thomas Jefferson had one built at Monticello somewhat later. Ice shipments to Charleston and other warmer American ports began soon after the Revolution; the export trade began in 1806 when Frederic Tudor shipped a load from Boston to Martinique. Tudor continuously shipped ice until about 1827, erecting large icehouses both in New England and, after 1816, at his ports of destination. Tudor became an ice tycoon as he adopted new methodologies for insulation, shipment, and even Nathan Wyeth's horse-drawn ice cutter. Ultimately he was able to reduce his waste from more than two-thirds to about 8 percent. Total natural-ice shipments from Boston, the largest export center, increased dramatically throughout the nineteenth century from 1,911 tons in 1827 to 43,125 in 1848 and 97,211 in 1860.

Ice prices declined with increasing competition, though the development of commercial and residential iceboxes after 1850 fueled continuing market expansion. The first cold-storage warehouse patent was issued to Benjamin Nyee of Decatur County, Illinois, in 1858. Greed, however, kept him from profiting from the offers that poured in to utilize his patent. Choosing to build his own warehouses, Nyee refused licensing offers from other entrepreneurs, including one for $100,000 for the right to build warehouses in New York City and another for $250,000 for a franchise for the state of Louisiana.

Lacking the connections and operating skills to be successful, he failed, and his patent was soon superseded by more efficient designs. A warehouse at New York's Fulton Market in 1865, for example, used a system of galvanized iron tanks filled with salt and ice to freeze large quantities of meat and game. Though its developers attempted to keep the process secret, warehouses using similar systems of tanks appeared around the country within a few years.

Ultimately the creation of the system of urban refrigerated warehouses set the stage for the widespread shipment of perishable foods from low-cost or off-season production areas to distant cities. The first refrigerated shipment of produce was begun by D. W. Davis, a local promoter, and a group of Cobden, Illinois, farmers in 1836 when they loaded a primitive refrigerator boxcar with strawberries and shipped them to Chicago. High freight costs and losses—when freight agents failed to properly re-ice shipments—combined to force the farmers to abandon the experiment after a short period. D. W. Davis, however, continued to work on the development of effective refrigerated railcars, receiving his first patent on a new design in 1865. He shipped his first carload of strawberries in 1868 and shipped peaches soon after. Thomas and Earl of Chicago created the first private-car line for shipping produce in 1887 and had 600 cars operating by 1891. It is estimated that there were 60,000 refrigerated railcars shipping produce, meat, and other perishable products in 1901.

Mechanical Refrigeration

William Cullen discovered that the heat exchange created during compression could be harnessed to create mechanical refrigeration in 1755. He abandoned further research after he found no practical use for his discovery, but others were not so easily discouraged. Experimentation continued on mechanical refrigeration as more efficient methods of compression were developed as well as methods of utilizing the heat exchange that took place when a volatile gas was transformed to liquid and vice-versa. The first commercial mechanical refrigeration unit was installed in a Louisiana warehouse in 1868. Three additional machines were installed the next year—all in the South.

Ice manufacturing continued to be concentrated in the South for more than a decade. The 1880 census shows twenty-nine plants in the South, including the largest in the nation, which had a daily production rate of 118 tons. Ice-cooled storage warehouses throughout the nation were soon switching to artificial cooling equipment; the days of the ice-cooled warehouses were numbered. By 1901 the nation could boast of 600 mechanically refrigerated warehouses with a capacity of 150 million cubic feet. This intense competition soon brought a decline to the natural-ice industry. The output of Hudson River natural-ice shippers, one of the nation's largest centers, declined by more than a quarter between 1904 and 1914.

Commercial Canning

The possibility of preserving cooked food by placing it in an airtight vessel to halt oxidization was first recognized by an Italian count, but he did not pursue his idea. It was not until Napoleon Bonaparte became frustrated with provisioning his armies and had the Directory of France offer a 12,000 franc prize to anyone who could create a foolproof method of preserving foods that a practical application of this concept was developed in 1795. Nicholas Appert, a sometime chef, pickler, and brewer, began experimenting with placing food in glass jars, heating them, and sealing them with a cork. Hundreds of exploding jars later he created an effective system of preserving food. He submitted the plan in 1809 and claimed the prize. Although his system was effective, it was not practical for mass production until an inexpensive method of manufacturing sealable containers was developed. Englishman Peter Durand (1810) contributed the containers, tin-coated steel cylinders with soldered caps at both ends. Unfortunately, these canisters had to be made by hand, making them relatively expensive to manufacture.

William Underwood brought the concept of canning foods commercially to America in 1817 and began packing a variety of fruits, pickles, and condiments in glass containers in 1819. He started utilizing an improved tin canister about 1840. Underwood is most famous today, however, for producing the canned brand-name product that was the most continuously packed in the nation over the longest period of time—Underwood's Deviled Ham. Unlike Underwood, most canners during this period concentrated on high-value products—lobsters, oysters, and salmon—because of the expense of hand-soldering every lid after filling. A system utilizing a small soldered cap was also introduced about this time, but it only slightly reduced production costs.

Costs finally began dropping after 1874 with the introduction of the closed steam pressure retort, which shortened cooking times by increasing cooking temperatures to 500 degrees. Continuous soldering machines raised worker productivity to more than 1,500 cans per day. Can costs plummeted in 1910 with the introduction of the double-crimped can, still commonly used today. This innovation allowed individual operators to manufacture as many as 35,000 cans per day. Reduced production costs combined with increased consumption of produce led to a rapid expansion of the canning industry into fruits and vegetables.

Farmers in specialty fruit and vegetable areas were plagued with excess production because it was difficult for them to estimate market demand with the complex distribution system in use at that time. Tomato plants growing in Florida might produce fruit all summer, but late-ripening produce close to the market forced harvesting to halt after a few weeks. Food processors soon realized that the construction of canneries in these chronic overproduction areas would allow them to process large quantities of excess, imperfect, and overripe produce at comparatively low costs. Canners also began contracting with farmers to harvest even greater quantities of produce, often ripening earlier or later in the season, to further lengthen the processing season.

Innovation soon also lowered preparation costs of produce for processing as well. The successful introduction of an automated corn-kernel remover in 1875 encouraged the development of a host of cost-saving devices. The first automated pea sheller (1883) replaced 600 seasonal employees at a single plant; similar savings were seen with the introduction of automated skinners, pitters, and similar equipment. Despite these savings, production costs for large and smaller manufacturers remained about the same because of the continued use of individual cooking kettles. It was not until after World War II that the combination of automated, continuous cookers; strong consumer brand identification; and self-service grocery stores allowed the largest competitors to drive smaller, local competitors from the marketplace.

Commercial canning was first concentrated in Maryland and Maine, where large harvests of lobsters and oysters kept processors busy part of each year. Processors soon realized that these plants could be used to process other foods. Each area operated with two seasons. Baltimore canneries packed oysters from September to June and packed tomatoes and other vegetables during the hot summer season when oysters were considered to be unsafe. Similarly, Maine canneries concentrated on lobster during the fall and spring (winter was too cold for any product to be available in volume) and turned their attention to sweet corn and wild blueberries during the summer season. The first commercial cannery on the West Coast opened in 1862 when one of these Maine competitors set up a plant on the Sacramento River to process the salmon harvest. A second plant was built on the Columbia River in Oregon soon afterward. It is not too surprising that the two largest canning companies in the nation in the 1914 census were the Alaska Packing Company, which processed salmon, and the California Fruit Canners Association (later to become Del Monte) in California. The CFCA packed 50,000 cans of produce in 1913; three-quarters of this was from California.

The National Canners Association listed 2,412 plants in its directory for 1915. Virginia and Maryland (405 canneries together), upstate New York, and New Jersey were the leading states, though the success of H. J. Heinz and other midwestern competitors soon brought a westward shift as new plants were built in Ohio, Indiana, Michigan, and Illinois. Companies in south-central Wisconsin soon became successful in processing cool-weather vegetables, and canning in the Minnesota River valley around LeSeur expanded rapidly with the success of Green Giant brand pea and corn products. The dominance of California came somewhat later, and initially most West Coast processors also had plants in the east.

Home Canning

It is impossible to estimate the total consumption of canned foods because of the continuing role of home canning in small towns and rural areas. John Mason created a glass jar with a sealable lid and metal ring that ultimately led the way to more varied diets for rural families without the cost of purchasing expensive canned goods. It is not known how many jars Mason actually manufactured, but the Ball brothers of Muncie, Indiana, were largely responsible for the rapid expansion of this industry after they entered the food-preservation glass-jar business in 1880. Like cans, these jars were hand blown into a mold and were initially quite expensive. Semiautomatic equipment was introduced to the United States in 1896, and the Ball brothers installed the equipment in 1898. They soon dominated the business. The development of fully automatic equipment after the turn of the century increased production to 3,500 jars per worker day.

Ball, Kerr, and the other major jar manufacturers remained private companies until after World War II, and little is known about total annual production. The government did tabulate 1.4 million jars for home canning in 1928, and production varied in that range until World War II food shortages brought renewed interest in home canning. More than 4.4 million jars were manufactured in 1943. Nineteen million lids were sold in 1947. Sales dropped to about 8 million lids in 1965 and then climbed to 21 million in 1976. Home canning declined in the 1980s but stabilized in the 1990s. The Ball Corporation created the independent Alltrista Corporation in 1993 to take over its home-canning manufacturing activities. Alltrista bought the Kerr home-canning manufacturing facilities and assets in 1996 and has become the largest producer of home-canning supplies in the United States and Canada. Little is known about how many jars or lids are manufactured annually because of Alltrista's dominance of the industry. Many of those who would have canned produce, jams and jellies, and meat products in the past now freeze them.

Although home canning is perceived by many to focus on jams, jellies, and fruit, actually almost anything could be and was preserved by home canners. In my family, for example, all of the aunts gathered at my grandmother's home and spent a day canning dozens of quarts of venison mincemeat from a secret family recipe. My wife's family canned more than 500 quarts of tomato puree each year just for the preparation of spaghetti sauce, as well as dozens of quarts of peaches and string beans. These amounts were typical throughout the nation, especially where publicly owned commercial-style canneries designed for small batches were established through government assistance to improve local diets.

Freezing

Frozen foods have been consumed for centuries, and many nineteenth-century cookbooks included a few words of advice on how to properly thaw meats that had become frozen. The commercial freezing of foods to preserve them, however, is a twentieth-century innovation. Clarence Birdseye spent the winter of 1915 working in Labrador, where he noted that meat and fish frozen quickly in the extreme winter temperatures could be thawed and consumed with no apparent loss of quality. Returning to the United States, he began developing an effective process for commercially freezing fish. He formed the General Sea Foods Company and began marketing frozen fish in 1925. Unfortunately, freezer cases were virtually unknown in grocery stores at this time, and grocers did not see any reason to install them. Retail consumers too were reluctant to purchase frozen fish because they had no way to store it in their homes. By 1928 the renamed General Foods Company had over 1.5 million pounds of frozen fish in storage and few customers.

Birdseye's partners lost heart and sold the company to the Postum Cereal Company in 1929. Post applied the newly acquired General Foods Company name to the combined corporation and renamed the frozen-product operation the Birdseye Frozen Food Division. Birdseye remained with the new company and was immediately assigned the task of creating a comprehensive line of frozen products. Sixteen different poultry, meat, fruit, and vegetable products were created in the first year.

The problem of retailing the frozen products remained. There were only about 500 retail stores with freezer space for selling frozen foods as late as 1933, and almost all of those were in New England. Making matters worse, Birdseye's original company had licensed a number of other processors to use General Foods patents to manufacture frozen foods under their own labels. Many of the products from these companies were of poor quality, and the public was beginning to be wary of frozen foods. Tapping the strong cash flow of the cereal operation, General Foods began repurchasing these licenses to help ensure future product quality while leasing freezer display cases to grocers at low cost. Concerned about the quality of its fruit and vegetable products, the company built mobile freezing units that could be moved from harvest region to harvest region to reduce the time between harvest and packaging. Heavy advertising, promotions, and the marketing clout of General Foods increased outlets for frozen foods to more than 10,000 by 1940.

World War II dramatically changed the market for frozen foods. The military placed orders for 70 million pounds of frozen foods at the outset of the war. These orders financed rapid improvements in processing methods and expansion of capacity for those companies who had been able to garner a share of this market. Although many competing food processors languished during the war years, frozen-food processors received a second bonus from the government in 1944, when frozen foods became unrationed at the same time that fresh and canned produce was still in short supply. At war's end General Foods, Swanson's, and a handful of other competitors were poised to dominate the infant industry, though it must be noted that frozen-food sales were still less than six pounds per capita in 1950.

Swanson's provided the next great innovation in frozen food. In 1899 Carl Swanson emigrated to Omaha, Nebraska, where he went into business with two friends, handling wholesale eggs, milk, butter, and poultry. By 1949 he operated the nation's largest distributorship for butter and eggs, grossing more than $50 million per year. The company had begun a successful frozen-chicken line during the war; his sons expanded this operation in 1951 to include the nation's first frozen prepared products—chicken, beef, and turkey pot pies. These pot pies were the right product at the right time. Their instant success was marred only by the many complaints that a single pie did not completely satisfy many customers. The brothers turned to their bacteriologist, Betty Cronin, who set about creating multicourse meals in a tray. The company created two new products in 1954 that included a turkey or pot roast entrée, mashed potatoes and gravy, and green peas. These new TV dinners, named after the new phenomenon sweeping the nation, were an instant hit. A fried-chicken dinner was added in 1955. Other entrepreneurs created a lightweight metal foldable table, called a TV tray, to allow the family to migrate from the kitchen table to the living room to eat their dinners while watching television. The dinner hour for millions of Americans was irrevocably changed.

Consolidation and Concentration of the Food Industry

The expanding market and improved technologies in the food industry coupled with the general business tenor of the time fostered the development of ever larger corporations hoping to increase profits by lowering costs through economies of scale. The beginnings of this process reach back to the 1820s and 1830s, but it did not become a dominant theme in the food industry until the 1880s. The Sugar Trust was possibly the largest single attempt to control production and marketing of a food product. The Sugar Refineries Company (later the American Sugar Company) was created in 1887 through the consolidation of eight refining corporations. Additional firms joined the cartel as the company gained control of 90 percent of the nation's sugar-refining capacity by 1892. The federal government challenged the cartel after the turn of the century, finally obtaining a consent decree from American Sugar in 1921, though competition had already eroded its market share to 24 percent.

No other single company in the food industry was ever able to obtain such clout in the marketplace, but significant concentration of the flour, meat, dairy, and produce industries did take place. As with all such processes, there were both positive and negative effects. The economies of scale and aggressive marketing dramatically lowered both producer and consumer costs while encouraging technological innovation. This same aggressive behavior also brought the specter of market manipulation, price fixing, and unfair business practices into the food industry. The evolution of these four food industries shaped the nation's food supply and ultimately the American diet. Understanding their history is crucial in understanding the transformation of the American diet.

Flour Milling

Declining freight rates brought by the completion of the Erie Canal and the later development of the rail transport system fostered the expansion of commercial wheat production into the Midwest after 1821. High-cost eastern producers were soon forced to find other crops as low-cost midwestern wheat flooded their traditional markets. Consumer wheat prices plummeted, making wheat flour more affordable to millions of Americans and ultimately creating an increased market demand for wheat breads. American wheat production exploded from 85 million bushels in 1839 to a half-billion bushels in 1880, setting the stage for a complete restructuring of the flour-milling industry. The center of the milling industry moved westward much more slowly than grain production, partially because eastern wheat farmers continued to play an important role in national output for many years and partially because mills tended to be located at transportation transaction points, not at production sites. New York continued as the leading flour-producing state in 1870, followed by Pennsylvania and Illinois. The last great East Coast mill, using twenty grinding stones, was built in Newark, New Jersey, in 1866. This 2,000-barrel-per-day behemoth was the largest in the nation but in a sense was doomed from the outset because millers in the West were already constructing even larger facilities with the more efficient roller technology.

St. Louis became the largest western milling center after the Civil War because of the high-quality winter wheat grown in Kansas and Nebraska. The situation quickly changed after 1870, when two French millers intro-

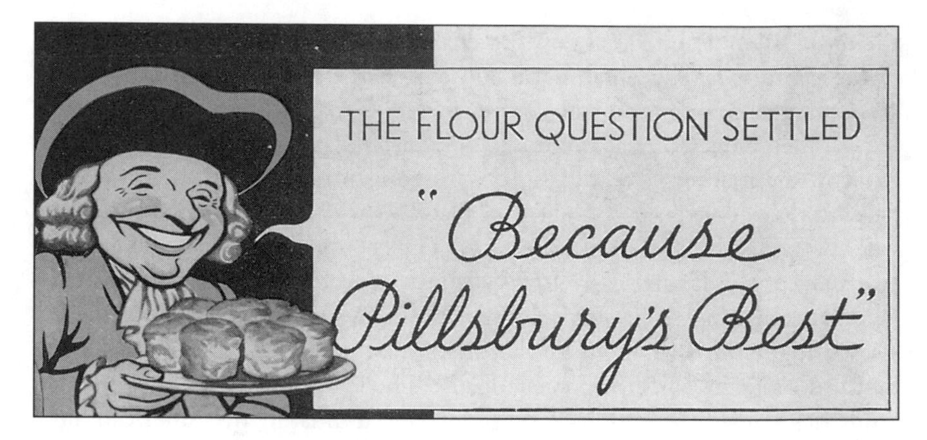

Strong and personal competition between Minneapolis millers Washburn (later General Mills) and Pillsbury quickly increased annual production to over 5 million barrels per year by 1884 and lowered prices, making Minneapolis the world's largest center of flour milling. (The Pillsbury Company)

duced a new, more efficient, European process to southern Minnesota and adjacent Nebraska. C. C. Washburn, a Minneapolis miller, hired their mill manager to install similar equipment in his Minneapolis mill. C. A. Pillsbury, his archrival, soon did the same. Adding the Hungarian process of using chilled iron and porcelain rollers to mill wheat, the Minneapolis competitors of Washburn and Pillsbury leapt over their competition a few years later to make Minneapolis the dominant milling center in the nation. Flour production in Minneapolis increased from 3.1 million barrels in 1881 to 5.1 million barrels in 1884.

Increased wheat production coupled with the development of more efficient milling operations had a tremendous effect on prices. Wheat-flour production rapidly expanded during the nineteenth century from about 40 million barrels in 1860 to 83 million in 1890 to almost 120 million on the eve of World War I. The average price of a 196-pound barrel was $10.03 in 1800, $5.55 in 1850, and $3.39 in 1900. Even with the modest inflation rate of the time, the relative cost of flour in 1900 was less than 20 percent of its 1800 cost.

Wheat-flour consumption predictably increased to over 200 pounds per person by 1900; cornmeal consumption declined to about 50 pounds, most of that in the American South and Appalachia. The factors underlying rising wheat-flour consumption were certainly more complex than just freight rates, manufacturing efficiencies, and lower retail prices. The arrival of more than a million new immigrants who were unfamiliar with cornbread also played a role. Changes in the kitchen, including the arrival of cast-iron stoves with easily controllable ovens, the rise of commercial bakeries, and the development of a system of grocery retail stores that promoted the purchase of these prepared foods were also potent forces in this change, but ultimately the declining price of wheat flour made the other factors possible.

The Meat Industry

Most Americans raised, butchered, and cured their own meats throughout the colonial period. Some butchers did slaughter the animals of others for a fee, but it was not until 1756 that the first permanent slaughter yard was opened in Boston. This meat was prepared primarily for fresh consumption. The first large-scale packing house had to wait a few years for William Pynchon of Springfield, Massachusetts, who cured and packed pork, bacon, and hams, primarily for the Caribbean trade. New England soon became the most important center for meatpacking, partially because of the region's access to high-quality Caribbean salt. Soon Connecticut ham became so famous that packers as far away as Pennsylvania began selling their products as being from Connecticut. Charleston became the largest center in the South with Cheraw bacon bringing good prices wherever it was sold.

Packing houses soon began appearing in the West as well. Many midwestern locations developed important industries, but it was the establishment of a packing plant in Cincinnati, Ohio, in 1818 that changed the course of the industry. By 1828 the Queen City began calling itself Porkopolis. It claimed that it had originated "the perfect way to pack fifteen bushels of corn into a pig and the pig into a barrel to be shipped throughout the world to feed mankind." By 1848 the city was packing 350,000 pigs during the slaughter season from late October to March.

Cincinnati became the largest center for pork packing in the world, spawning a variety of by-product manufacturers as well. Many of the unusable parts were simply thrown off the wharves into the Ohio River; local residents had the greatest supply of inexpensive fresh meat in the world from cuts that did not lend themselves to salting but were popular when cooked fresh. Other by-products, such as lard and tallow, were sold to companies that processed them into glue, soap, and other products. Proctor and Gamble, probably the largest of these companies, survived the decline of the packing industry and the use of lard and tallow in soap products to remain one of the world's largest manufacturers of soaps and personal care products.

Virtually all the animals slaughtered in Cincinnati were cut into about eight-pound pieces, pickled, and then packed in barrels with salt. The lack of refrigeration meant that the packing houses opened after the first hard freeze and closed for the season as soon as warm weather became prevalent. Several made attempts to refrigerate storage facilities, but packers generally were unable to alter the packing season until late in the century.

Other smaller meatpacking centers also developed during this period, including Bardstown and Louisville, Kentucky, Chicago, and Milwaukee, but it was the development of a regional rail network focusing on Chicago that allowed it to challenge Porkopolis as the center of meatpacking. The firm of Clyburn and Dole made the first meat shipment from Chicago in 1832. By 1880 more than 9 million cattle, hogs, and sheep were being sent annually to the Chicago stockyards for slaughter. The city not only had become the "hog butcher of the world" in the 1870s, it began to play an instrumental role in reshaping the nation's meat-eating habits.

The key to Chicago's dominance of the meat industry was the development of the Union Stockyard. The livestock business grew rapidly during the 1860s as the city expanded its iron tentacles throughout the midwestern agricultural heartland. A central gathering point for cattle shipped eastward from the Great Plains, Chicago faced a seemingly insoluble problem, brought on by its own success. The eastern and western rail lines did not meet at a central staging area, and the cattle were herded down city streets, mooing and defecating. The cattle drovers were always impatient to make their sales and get away from the city as soon as possible. The buyers and packers were frustrated at not being able to easily move between the railyards of the nine rail companies that serviced the city to purchase the cattle that they needed. The residents were upset about the noise and smell. Finally, on one especially difficult day several cattle drovers ignored warning signals that the Rush Street drawbridge was opening. They continued herding their cattle onto the bridge, and as the bridge began to open, even more animals rushed out onto the now-unsupported span. The unbalanced bridge slowly twisted and collapsed under the weight, dropping pedestrians and cattle into the cold Chicago River. A dozen pedestrians and more than fifty cattle drowned in the turmoil of sinking bridge, thrashing cattle, and screaming people, causing an instant uproar in the city over the practice of herding cattle through city streets.

A plan was immediately drawn up for a single, great, self-contained stockyard accessible to all rail companies, buyers, sellers, and packing companies. Initially working with more than a half square mile of land, the organizers developed more than thirty miles of drainage canals and fifteen miles of track, holding pens, unloading facilities, offices, and a hotel. The yards grew to many times their original size; the volume grew to more than 13.5 million

TABLE 3.1 Commercial Meatpacking: 1890

Type	*Pounds Sold*
Beef, sold fresh	2,708,319,960
Beef, canned	133,428,456
Beef, salted or cured	576,289,731
Pork, sold fresh	1,125,648,541
Pork, salt	1,264,956,237
Hams	529,387,213
Bacon	666,229,376
Sausage	149,281,545
Mutton, sold fresh	267,353,788

SOURCE: *Eleventh Census of the United States, Manufacturing*, 1893.

animals per year (only 3.5 million cattle) in 1892. The power of the five Chicago-based meatpacking companies, Armour and Company, Swift and Company, Morris and Company, Wilson and Company, and Cudahy Packing Company, grew apace. These companies controlled the Chicago Union Stockyards through most of its history, and according to a report by the Federal Trade Commission, by the turn of the century they controlled the nation's meat supply as well.

Americans continued to primarily consume cured and salt pork throughout the nineteenth century. The fear of food contamination was high, knowledge of the decay process slight, and refrigeration virtually nonexistent. Change began during the Civil War. The War Department required Chicago meatpackers to include fresh beef in their shipments to the troops along the Mississippi. This introduced large numbers of men to fresh beef while forcing the meatpackers to develop facilities to ship fresh product.

Commercial meatpackers turned to fresh-meat production much earlier than home meatpackers. Commercial meatpackers shipped 2.7 billion pounds of fresh beef and 1.1 billion pounds of fresh pork in 1890, easily surpassing the production of preserved pork. Pork did continue to dominate total meat consumption with total shipments of 4.2 billion pounds that year (Table 3.1). Despite the common misconception that beef packing dominated the Chicago meatpacking industry, beef accounted for only about 30 percent of the meat shipped from the Chicago Union Stockyards in 1890.

Beef consumption increased in the early twentieth century to about equal that of pork before World War I. Consumption remained about equal throughout the first half of the twentieth century—for example, 71 pounds of beef versus 69 pounds of pork per capita in 1950. The nation's preference for pork rapidly changed after that time. By 1960 Americans were consum-

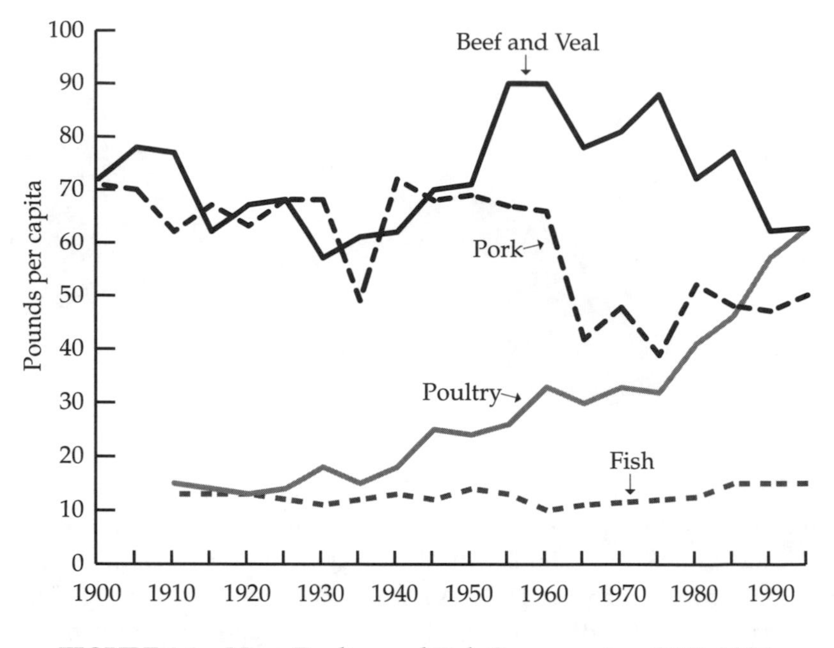

FIGURE 3.2 Meat, Poultry, and Fish Consumption: 1900–1994

ing 85 pounds of beef for every 65 pounds of pork, and by 1970 the ratio had changed to more than twice as much beef as pork. This was one of the most dramatic shifts in American food preferences in the entire history of the nation (see Figure 3.2).

It would be an attractive idea that Ray Kroc's purchase of the franchise rights to McDonald's in 1954 along with the appearance of Burger King in the same year was at the base of this change, but at best the rise of the fast-food hamburger is only a single factor in the changes that took place. The changing character of American family life also played an important role in the changing diet. Married women in the workforce increased from 5 million in 1940 to 9.3 million in 1950 to more than 13 million in 1960. About 80 percent of these households were dual–wage earner homes, and a good many had children. The rise of thousands of inexpensive restaurants, seemingly on every corner in suburbia; the instant popularity of hamburgers, french fries, and a soft drink among the children of these suburbanites; and the shift in family activity from patterns of unplanned play to character-building, adult-supervised sports (Little League and similar activities), lessons, and group activities meant that "Mom's taxi" was born. Mom balanced job and housework with shuffling the kids from athletic field to athletic field for after-school practices—all the while passing attractive, brightly lit McDonald's and Burger King restaurants. This lifestyle made the restaurateurs instant millionaires and inexorably contributed to the shift toward beef. It's amazing that the chains didn't sponsor more Little League teams than they did.

Despite the importance of chicken in our collective historical consciousness, the average American consumed only about 15 pounds of poultry per year even as late as 1910. This pattern remained stable through the depression but began changing during World War II. Dwindling supplies of pork and beef and the ease of raising chickens on the farm or even in town if the house lot was large enough raised per capita consumption to about 22 pounds by the end of the war. Chicken consumption jumped in the 1950s. Whether this is due to Col. Sanders's efforts at franchising his famous recipe, as the company would have you believe, is immaterial. The fast-food boom was upon us with chicken franchisers offering a popular alternative to the endless diet of hamburgers.

The health concerns about all meat consumption since the 1960s have played an important role in the continued rise of poultry consumption generally and the decline of both beef and pork consumption since the late 1970s. Chicken consumption passed pork consumption in the mid-1990s, and turkey consumption rose to about 15 pounds per capita. Today one finds turkey ham, turkey bacon, and a host of pseudopork products at every deli and sandwich counter. Seafood, almost universally believed to be the best protein choice of the four because of its lower fat content, has benefited little from the national health mania of the past twenty-five years. Total seafood consumption rose from a little over 10 pounds per person in 1960 to about 15 pounds today. Considering that a significant percentage of the fish consumed is fried, this slow transition is not all bad. The recent expansion of fish farming with heavy investments from three of the nation's largest protein companies, ConAgra, Tyson's, and Gold Kist, may well signal a future attempt by processors and retailers to convince consumers to eat more fish. Although fish farming has focused on catfish because of its high feed-to-meat ratio, there has also been a notable expansion of Atlantic salmon, tilapia, and clam farming in recent years.

The Dairy Industry

Most nineteenth-century American farmers maintained a milk cow or two to ensure sufficient supplies of cream, butter, cheese, and milk for household use. Farm wives continued utilizing milk in largely the same manner as their ancestors after the initial frontier period passed. Urban cooks and homemakers, however, were unable to obtain adequate supplies of dairy products. Faced with dropping grain prices because of inexpensive wheat shipped from the Midwest, some eastern farmers soon seized the opportunity to specialize their operations. Many New England and New York farmers outside the urban fringe initially shifted to wool production to supply the growing textile miles in southern New England, but increasing foreign competition and higher profits from the production of perishable foodstuffs for the urban market led many to shift to operating dairies, especially those farming in the cooler, more isolated New England and upstate New York areas.

New York was the leading dairy state in 1839, producing almost a third of the nation's commercial dairy output by value. Pennsylvania was the second most important state with only about one-third the production. New York herds reached a million head by 1860 as the nation's first milk shed evolved. The first experimental train to stop at rural milk "stations" made its first run from Chester (Orange County, New York) to New York City in 1842. It was an immediate success, and milk runs soon became a standard around all of the nation's larger cities.

Production moved westward as the midwestern economy matured. Iowa passed New York in numbers of dairy cows in 1890, with Wisconsin becoming the most important state a few years later. Western production tended to concentrate on solid products outside of the areas servicing the Chicago-Milwaukee urbanized zone; eastern production continued to be primarily focused on the fluid products.

The greatest problem facing the infant dairy industry at this time was not volume but quality. Illnesses caused by spoiled and contaminated food reached virtually epidemic levels among the urban poor and middle class in the late nineteenth century. Although there were problems with all foods becoming contaminated, milk and other dairy products were especially vulnerable. Each year, tens of thousands of children suffered from "milk sickness." Indeed, one wonders why so much spoiled milk was consumed given its distinctive smell and taste.

Home milk delivery began in the late nineteenth century to speed the transfer of milk through the distribution network, and the glass milk bottle was introduced in 1886 so that the consumer could inspect the product and see that the bottle was clean, if not sanitary. Pasteurization was introduced in 1895, but pasteurization changed the taste of the milk, and many refused to drink the new product—much like modern consumers have resisted sterilized milk because of its flavor change. Eastern consumers were especially reluctant to shift to pasteurized milk, and only about a third of milk sold in eastern urban markets was treated as late as 1912. In contrast, more than half of all milk sold in Chicago and three-quarters sold in Milwaukee was pasteurized at that time.

Government intervention in the dairy industry since that time has been unparalleled. Concerns have focused on the production of a hygienic product and the maintenance of low consumer prices to ensure the continuing availability of milk for children. Hygiene programs have focused on the dual issue of tuberculosis testing of herds and bacteria testing of fluid milk to control disease. Though not required, virtually all milk today is pasteurized to meet the cleanliness standards, though at least one European dairy company operating in the United States now offers sterilized milk as well. A complex program of federal and some state programs regulate consumer milk prices to maintain artificially low fluid milk prices. An outcome of this program has been the consolidation of the industry to reduce production costs. There are less than 1,000 milk handlers in the nation today; seven companies, three of them operated by grocery store chains, market 75 percent of all fluid milk.

Produce

Imported fruits and vegetables have almost always been a continuing part of American cuisine. Store accounts and import records suggest that even mundane products such as potatoes and yams, as well as tropical fruits, rice, and a host of other food products, were routinely imported to the United States throughout the colonial period. Amelia Simmons (1796) provides lengthy discussions about the selection of mangoes, Spanish potatoes (sweet potatoes), white potatoes, and other perishables in the late eighteenth century.

The first specialty production areas of perishables in the United States were clustered around the infant cities of the eastern seaboard, especially along the Hudson River, in northern and central New Jersey, southeastern Pennsylvania, and the eastern shore of Maryland and Virginia. Increasing market demand encouraged entrepreneurs to experiment with the production of off-season produce. Growers in Norfolk, Virginia, realized in the mid-nineteenth century that they could ship early spring tomatoes northward weeks before they were locally available in eastern markets. Soon producers on Johns Island near Charleston and Lady's Island near Beaufort, South Carolina, also joined this profitable trade.

The refrigerated railcar played an amazing role in restructuring the American specialty agricultural and produce industries. Though shipments of perishables were initiated by farmers and entrepreneurs, the rail companies themselves soon realized the potential of this market. The shipment of off-season produce from the West was obvious, but the Illinois Central and others also realized that they could take advantage of the north-south climatic banding by shipping perishables from the rural South as well. Arkansas quickly became one of the largest centers of apple production in the United States, and farmers along the rail lines in southern Illinois, Arkansas, and Mississippi were soon specializing in strawberries and other fragile crops for rail shipment to Chicago.

Some insight into this complex support system can be gained by looking at the structure of the Chicago Produce Market. Occupying a sprawling area along South Water Street and supporting more than 700 produce dealers, it established a ready market for perishables. Dealers purchased incoming produce in any volume, consolidated or broke their purchases into manageable units, and resold their purchases locally or in other markets in a matter of a few hours. The market's volume encouraged producers to expand commercial production in the knowledge that there would be a ready market for their output and established de facto wholesale prices for grocers and other processors.

A complex set of standing relationships evolved in Chicago and other major markets among growers, middlemen, and retailers. Essentially sales fell into four primary relationships. Traditional brokering, producers selling to wholesalers who resell to retailers, has typically been the most important method of distributing produce.

Growers' cooperatives began appearing in larger numbers in the 1890s as a means of giving farmers greater control over the sale and distribution of their products. The California Raisin Growers (Sunkist), the California Fruit Growers Exchange (selling their processed products under the Del Monte label), and Gold Kist are some of the more successful of these. These growers typically do their own consolidating prior to shipping and often handle sales and distribution as well.

Contract farming, wherein processors and distributors contract with farmers prior to the season for the production of a specific crop, is of increasing importance today. The terms of the purchase price are set at that time either for a fixed amount or for some percentage of the open market price of the crop at the time of harvest. This system guarantees processors and distributors of adequate supplies of product for sale when needed and guarantees farmers that there will be a market for their produce when it is harvested. Unfortunately, the farmer takes most of the risk and has little hope of large rewards if prices rise beyond expectations, though the ultimate criticism of the system is that it transforms the farmer from the role of entrepreneur to that of a wage laborer.

The Chicago Produce Market

South Water Street is a short east-and-west street which lies between the downtown business district ("the Loop") on the south and the Chicago River on the north. The portion used for produce-market purposes is a scant half-mile in length. . . . Generally speaking, fruit and vegetable dealers are located in the eastern part of the district while the western end contains the establishments which specialize in meat, poultry, and dairy products. Likewise, the initiated observe a distinction between the north and south sides of the street, the latter being known as the "busy side." It has the obvious advantages in summer of being the shady side, and stores on this side of the street run back to an alley, which is convenient for handling goods. Both sides of the street are lined with low brick stores, none of them new, and many of them dating to the days just following the great fire of '71. The ground floor of these buildings is used for storage and salesroom purposes. . . . There is commonly a cashier's booth at one side of the street floor, but the general offices are frequently located on the second floor. The basement and the remaining upper stories are used for storage, repacking fruits and vegetables, candling eggs, [and] dressing poultry. . . . If [the business] does not require so much space the basement is often rented to some other firm or individual (most commonly a dealer in bananas, butter, or eggs), and the upper floors [are] occupied by car-lot wholesalers, whose business does not require the use of salesrooms, or by dealers in feathers, pelts, old bags, and the like.

South Water Street is sometimes referred to as "the busiest street in the world." This, if true, is not entirely to its credit. Some of its "busyness" is mere wasted effort. The extraordinary mass of vehicles which choke this thoroughfare, the endless piles of crates and sacks and barrels which obstruct its sidewalks, the seething crowd which fills every inter-space until it impedes its own progress, give evidence magnificently of the large volume of business which centers in this district, but they also give evidence deplorably of the lack of up-to-date equipment for the handling of business. The number of teams on the street is an index not merely of the volume of traffic being handled, but also of the time wasted in the handling of it. . . .

The business of South Water Street overflows into the adjacent side streets [which are] utilized as a parking space for teams waiting to load up supplies purchased on "the Street." Some South Water Street merchants find themselves so fully occupied with shipments to out-of-town customers that the space in front of their places of business is entirely taken up by teams receiving their loads for the various express and freight depots. In such cases local buyers are compelled to wait "around the corner" until their wants can be supplied and their goods brought to them on a hand truck. . . .

(continues)

(continued)

Finally, the account of the central produce district will not be complete without mention of the many auction rooms and cold-storage warehouses, most of which are located just across the river. . . . There are two auction companies: their business is limited to the selling of citrus fruit from Florida and California and of deciduous fruit from California and the Northwest. The display and salesrooms of the Green Fruit Auction Company are located in a building which occupies the north bank of the river . . . , while the Central Fruit Auction Company occupies the eastern portion of the same building. . . .

Cold-storage warehouses are an important adjunct to the produce market. Even though the individual dealer has his private chill-room or may go so far as to have the larger part of his basement provided with refrigeration, the larger establishments must be depended upon when goods are to be stored in large quantities, for long times, or at the lowest temperatures. Naturally the cold-storage companies seek a location as close as possible to the market section. Several of the most important are . . . located two blocks north of the Water Street market. . . . There are also four cold-storage plants adjacent to the freight terminals . . . and quite a group of them in the Stockyards district. Most of these latter are controlled by the packers themselves, but, besides handling packing-house products, they rent space to produce dealers. This is mostly used for storing poultry, butter, and eggs. (Nourse, 1918, 15–19)

Consumption of canned fruits and vegetables grew rapidly during the late nineteenth century, though it was not until 1913 that American consumption of commercially canned fruits surpassed that of dried. Vegetable growers were the first beneficiaries of this revolution; Americans consumed more than fifteen pounds of commercially canned vegetables in 1910 and about a half-pound a week by 1925. Canned-fruit consumption was slower to rise, partially because the system of refrigerated railcars and mechanically refrigerated warehouses brought inexpensive fresh fruit to most cities.

Increasing produce consumption brought the consolidation of processors, the development of specialized farms producing fruits and vegetables solely for processing, and vast quantities of advertising specifically targeted at increasing consumption. The California Central Valley soon became the largest center of peach and tomato farming in the world yet shipped only a trickle to local markets as fresh produce. Virtually the entire crops were canned. Similarly, large quantities of plums and grapes in the Central Valley were dried before sale for prunes and raisins until the past decade or two.

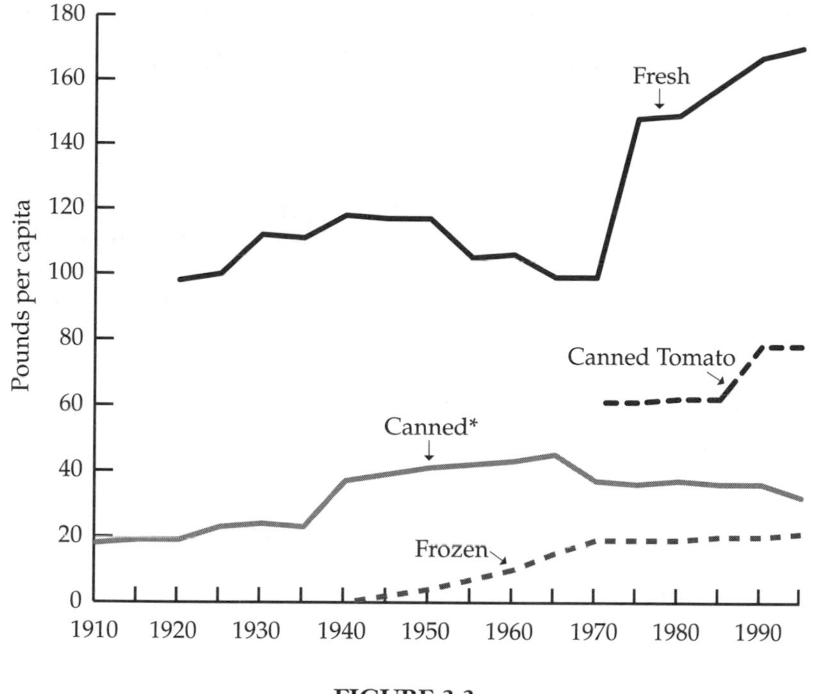

FIGURE 3.3
Consumption of Processed Vegetables: 1910–1994

Some vegetables have continued to be consumed in large quantities in their canned form to this day. Tomatoes are the most important canned vegetable; the average American consumes almost 70 pounds of canned tomato products per year.

Other important canned vegetables include sweet corn (10 pounds), pickles (4.3 pounds), snap beans (4 pounds), and peas (less than 2 pounds). Consumption of all canned vegetables, except tomato products, is declining today (Figure 3.3).

More drastic in changing the food supply has been the globalization of the food distribution system and the corporations that control it. Declining transportation costs and the creation of more sophisticated storage technologies have made it possible to move virtually any produce item from market to consumer at a reasonable price. On a tour through my favorite market during the winter, I find green peppers from Costa Rica, onions from Peru, peaches from Chile, and fresh ginger from the Fiji Islands. The American diet has become exceedingly complex in response to these changes.

The New Pantry: Last Thoughts

The monotony of the traditional nineteenth-century diet is difficult to imagine today as we visit our modern grocery stores filled to the ceilings with every possible food. New hybrids and new prepared foods and vast improvements in storage and transportation technologies have changed the American diet. These changes have in turn changed demand, and with changes in demand come changes in pricing and in distribution patterns. Even though many Americans may limit their diets to particular foods, they have come to expect that changes in personal preferences—whether brought about by lifestyle changes or by boredom or by changing health needs—will be met with availability. Availability, in turn, continually affects personal preferences.

4

Too Busy to Cook: The Coming of Prepared Foods

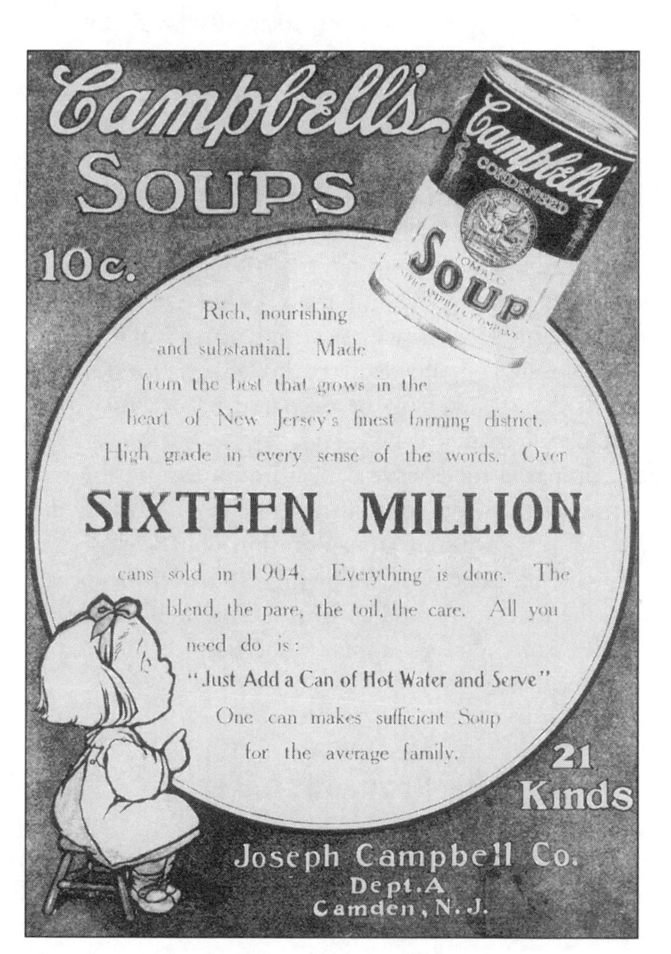

Nothing says lovin' like something from the oven.

—The Pillsbury Company

The American cook, whether housewife or hired girl, typically found little incentive to stray from traditional foods regardless of the new foods appearing in the local store. The introduction of increasing numbers of prepared dishes on grocers' shelves, however, brought whole new meals within the grasp of consumers without the commitment of learning how to cook in an entirely new manner. A little advertising, thousands of free samples, some cooking advice, and a few old-fashioned guilt trips about taking care of one's family soon made these foods an integral part of modern life.

Prepared foods are so pervasive in our society today that it is impossible to estimate their impact. Few home cooks bake a cake from plain flour, sugar, and baking powder, fewer prepare soup from scratch, and the typical side dish on the dinner table is most likely from a box or can. Meal replacement is the hottest new concept in the hospitality and supermarket industries; they recognize that the nation's beleaguered workers in dual-worker households are increasingly finding themselves too busy to shop and cook and too tired to eat out. Prepared foods play an even more pervasive role in the restaurant trade, where larger and larger numbers of stores are becoming little more than commercial dining rooms with assembly lines and a wait staff. Restaurants have largely relied on outside prepared desserts for decades; today even the signature dishes are likely to be prepared in a factory somewhere and shipped to site like so many frozen bricks.

Ultimately, prepared foods have had more impact on the way Americans dine than almost any other factor. No discussion of evolving dietary preferences can be complete without an exploration of at least some elements of the prepared-food industry. The following discussion on breakfast cereals, canned soups, the rise of the food conglomerates, and prepared foods in the restaurant trade is intended to capture something of the complexity of this pervasive industry.

The Breakfast Derby

There has never been a standard breakfast in the United States, even though the so-called American breakfast of eggs, hash browns, and toast is almost universally perceived to be the nation's preferred meal by the hospitality in-

Half of all Americans start each day with a bowl of cereal. (Sandra Weinwurm)

dustry here and abroad. Indeed, that meal is as close as anything to a national norm when Americans dine in restaurants, but home meals have always tended to be a bit different.

Half of all Americans start each day with a bowl of cereal, consuming 160 bowls of cereal per capita per year. Americans purchase almost 3 billion packages of breakfast cereal each year, and those packages contain almost a billion pounds of sweeteners (Bruce and Crawford, 1995). Breakfast cereals are the third most popular product sold in supermarkets, following carbonated beverages and milk but ahead of bread. The development of brand-name breakfast cereals in the late nineteenth century represents one of the greatest financial successes in the prepared-food industry as well as one of the most dramatic alterations of a culture's dietary preferences.

The first prepared breakfast cereal, Granula, was created by Dr. James Caleb Jackson in 1863. This new cereal was prepared from a simple dough of graham flour and water that was baked into a brick, broken into berry-sized granules, and then rebaked. The final product was brick hard and had to be soaked in milk for hours before it could be eaten. Dr. Jackson created Granula as a part of his hydropathic health treatment program based on exercise, fresh air, an abundance of water both for drinking and bathing, and a healthy diet. Neither Granula nor its companion health "coffee," Somo, caught the public imagination, but they laid the foundations for the nineteenth-century health food revolution.

The health food crusade remained strong throughout the nineteenth century with a small band of vocal followers. The appearance of two new religious cults, the Seventh Day Adventists and Church of the Latter Day Saints of Jesus Christ, gave the movement even more followers, as both had dietary restrictions as a part of their religious regime. The Seventh Day Adventists settled in Battle Creek, Michigan, where Brother and Sister White led the faithful. In 1863 Sister White received a series of visions specifying a healthy dietary regime; these were followed a few years later by instructions to found the Western Health Reform Institute. John Harvey Kellogg worked closely with the Whites as a boy, and they encouraged him to pursue an education in medicine. Kellogg became a noted surgeon and took over management of the Western Institute, which he renamed the Medical and Surgical Sanitarium, in 1876. Life at the sanitarium was chronicled hilariously in a book and a movie, both entitled *The Road to Wellville*, but for Kellogg and his followers, taking care of one's body was serious business. Patients and visitors included many of the most important people of the day, ranging from William Howard Taft to Billy Sunday and J. C. Penney to Johnny Weismuller. Whereas Dr. Kellogg's enema and sexual abstinence cures may seem a bit flaky today, they were better than many of the alternatives and certainly held less potential for harm to the patient than many others of the time.

Kellogg strongly believed in the power of bran. He began experimenting with the creation of a cereal in the 1870s, introducing his own Granula in 1880. Dr. Jackson successfully sued him in 1881, and the cereal was renamed Granola. Kellogg was selling more than 100 tons a year of this concoction by the end of the decade. More important, Kellogg began experimenting with the creation of a flaked cereal. Though accounts differ, John Harvey Kellogg and his brother, William Keith, began experimenting with the creation of wheat flakes, which they patented in 1894. Granose Flakes hit the market in 1896 and became an instant commercial success.

Sanitas Toasted Corn Flakes followed in 1898. The new product was successful, though W. K. Kellogg felt that its flat flavor inhibited sales. John Harvey took a lengthy tour of Europe in 1903, and William Keith took this opportunity to begin experimenting with the cereal's flavor. Discovering that a small amount of sugar made a great difference in flavor, he added the new ingredient to the manufacturing formula immediately. The new sweetened flakes were an instant success, and orders increased dramatically. John Harvey was furious when he returned to discover that this health product had been adulterated. The sanitarium board backed his brother, and the sweetened product continued to be sold. W. K. immediately launched a successful national advertising campaign for Toasted Corn Flakes, "the breakfast treat that makes you eat."

Stress between the brothers was so intense that William Keith Kellogg finally obtained a partner and convinced the sanitarium board to sell him all rights to the corn flake product. The independent Battle Creek Toasted Corn Flake Company (1906) was selling 2,900 cases of corn flakes a day by late 1906. By 1911 corn flakes had become so popular that 106 competitors had appeared on the scene in the United States alone.

Kellogg's greatest competitor in the cereal business was C. W. Post. Post had originally come to the sanitarium in 1891 as a patient with a severe digestion problem. Kellogg was unable to help him, but Post became fascinated with the ersatz coffee served at the facility. Running out of money, Post moved in with his sister's cousin in Battle Creek. He continued his coffee therapy at her home and underwent a miraculous recovery. Convinced that the cereal coffee had cured him, he approached John Harvey Kellogg with an offer to market the product. Rebuffed, he was undaunted. Founding his own health spa, La Vita Inn, he continued research on the ersatz coffee in his kitchen, ultimately selecting a recipe containing 22.5 percent wheat berries, 10 percent molasses, and 67.5 percent wheat bran. He named the new product Postum and began a strong marketing campaign in 1895. Sales exceeded a quarter million dollars annually within two years.

Post used the strong cash flow from Postum to create a new cereal product of whole wheat, malted barley, and yeast, which he called Grape-Nuts (1898) for its nutty flavor. Elijah's Manna, a corn flake cereal, followed in 1906. The religious community was outraged at the name, and the product was banned in England. Post renamed the product Post Toasties and in 1908 replaced the picture of Elijah in the desert with Cinderella sitting by the fireside. Post Toasties became the company's most profitable product with a reported 1908 profit of $2.2 million versus $1.7 million from Grape-Nuts and $1.4 million from Postum. Post never saw his ultimate success over the hated Kellogg brothers, as he committed suicide while in postoperative depression four years later, well before his company flowered to its ultimate extent.

The development of the hot (ready-to-cook) breakfast cereal industry was also moving apace with its cold (ready-to-eat) cereal competitors. Pre-Neolithic foragers cooked grains by parching them or cooking them as a gruel of grain and water. Cooked grain, porridge in England, was a common dish throughout medieval Europe and continued to be an important dietary staple among the European immigrants in America. The transformation of this simple dish into a brand-name prepared food, however, did not take place until the late nineteenth century.

The American oatmeal breakfast cereal business began soon after Ferdinand Schumacher emigrated to Akron, Ohio, from his home in Germany. Starting a grocery, Schumacher was surprised to discover that oatmeal was not an important food in this country. Seeing an opportunity to expand sales, he purchased a grinder and began preparing his own oatmeal product to sell in his store. Within a few years he was selling 20 barrels of oatmeal a day to customers throughout the eastern United States, and he closed his store. Soon dubbed the oatmeal king, Schumacher raised production to over 2,000 barrels a day in the 1870s.

Success breeds competition, and oatmeal was no different than any other product. Henry Crowell, a marketing genius of the day, purchased the bankrupt Quaker Oats Company in the 1870s and was soon Schumacher's largest competitor. He quickly recognized that although the three largest competitors were selling more and more oatmeal, their profits were stagnant. He approached his two largest competitors with the idea of a merger, and the American Cereal Company was born. The new company, renamed Quaker Oats in 1906, became and still remains the dominant oatmeal producer in the world.

The success of Quaker Oats spawned a covey of cooked-cereal imitators. The largest of these was created in 1893 when Thomson Amidon, head miller of the failing Diamond Milling Company, discovered that the middlings left over from the wheat-milling process could be cooked into a dandy breakfast cereal. He convinced the owners to make one last attempt to save the company by selling the new product. Dubbed Cream of Wheat for its white color, it had a simple package design that prominently displayed a smiling African American chef, primarily because the printer found a woodblock in his storeroom with that image. Samples were created, and a case was sent to New York as part of a carload shipment of flour. Three hours after Cream of Wheat arrived in New York, a telegram arrived at the South Dakota mill reading, "Forget the flour. Send us a car of CREAM OF WHEAT."

Puffed cereals were probably the most unique of the breakfast products of that period. Alexander Anderson, a chemist, became interested in the potential of exploding or puffing high-moisture grains as a breakfast cereal. Seven frustrating years later he presented his laboratory success to Henry Crowell and convinced him to finance the development of a commercial process. He employed a breech-loading cannon at one point to produce the "cereal shot from cannons." Quaker Oats sold 200,000 boxes of the new cereal its first year, 1904, and 300,000 boxes in the second.

All of the cereal companies employed massive advertising campaigns. One of the classic examples involved the *Sergeant Preston of the Yukon* radio series in 1955. The cereal company obtained a plot of land in Canada's Yukon Territory and promised purchasers a deed for one square inch of the Yukon for every box purchased. Children, including yours truly, pestered their parents to purchase a box of the cereal so they could own goldlands in the Yukon. Thirty-five million boxes of puffed cereals were sold in weeks, though at least one box sat unconsumed in a cupboard until it was recycled several years later. As a side note, a group of deed holders acquired 10,800 of these "deeds" in the 1960s and attempted to claim the land. The land the company had acquired, however, had reverted back to the Canadian government for nonpayment of taxes, and the claimants were stymied.

The prepared-breakfast-food industry was born and grew into a monster in only two decades. Several hundred competitors jockeyed for market share in a crowded field where there had been barely a concept just a few years previous. Four companies dominated—Kellogg's, Post, Quaker Oats, and Cream of Wheat—but others followed, coming and going as the market continued to shift. Though the story is interesting in itself, one must wonder why the American public adopted this radically new approach to breakfast in such a short period. Membership in the Seventh Day Adventist and Mormon churches combined never reached more than a few percent of the population. Even a cursory examination of the current eating habits of the population demonstrates that nutrition or supposed health qualities play only a small role in the decision process.

Convenience is the plain and simple answer. The ratio of rural to urban residents in America was transformed just after 1900 as more Americans became urban than rural. The 1900 census found more than 6 million living in cities of a million and 14 million living in cities over 100,000 in population. Rural people could and did eat prepared cereals, but an urban housewife would have had to rise very early to prepare the children for school as well as provide a hearty, traditional breakfast: Both husband and children left early for job and school, and a woodstove might take a half-hour to heat up sufficiently to cook biscuits.

Early prepared-cereal consumers were a small percentage of the total potential market but still represented a massive financial lode to be mined by these early food entrepreneurs. Ready-to-cook cereals, however, haven't grown in concert with the ready-to-eat varieties. The consumption of ready-to-eat cereal jumped from about 7 pounds per capita soon after World War

II to 15.5 pounds per capita in 1994. Simultaneously, consumption of the traditional breakfast food—sausage, bacon, and eggs—has dropped precipitously, especially since 1950. Less than 10 percent of Americans today consume these items at breakfast on any given day, and most consumers who do are over forty years of age.

The traditional breakfast favorites have not entirely been replaced by prepared breakfast cereals. Americans eat breakfast away from home far more than ever before, and they tend to order some form of eggs, traditional breakfast meats, and pancakes. The success of McDonald's and Hardee's in providing inexpensive breakfast meals in the 1970s began pulling many middle- and lower-income people away from their home breakfast tables; the more recent expansion of doughnut shop chains, the arrival of hundreds of coffee houses (which also serve bakery items for breakfast), and the chaining of the bagel shop have all had their impact on the American breakfast. The prepared-food manufacturers too have entered the fray with other more convenient products. The Kellogg Cereal Company introduced the Pop Tart in 1964 and the Eggo frozen waffle in 1970. A walk down the grocery aisle today reveals waffles, sausage biscuits, eggs, and even breakfast burritos side by side in the freezer cases. The ultimate key to the change remains the same—convenience. As more and more households have less and less time for food preparation, breakfast has become a meal that is easily shifted to the shoulders of prepared-food manufacturers or commercial establishments. The cost of eating at a fast-food chain or diner is so small that for most of the middle class, the time lost in purchasing the foods, their preparation, and cleanup outweighs the actual money spent. Weekends, in urban areas especially, find even more millions of Americans brunching with friends or family in a relaxed restaurant atmosphere with little incentive to cook complex meals or wash dishes.

From Soup to . . .

The development of more automated machinery and less expensive containers during the 1890s led to the establishment of hundreds of prepared-food manufacturing companies; most of the nation's greatest prepared-food companies date from this era. Many of these companies focused on the preparation of traditional products that would ease the housewife's workload, such as beef stew, mincemeat, ketchup (at one point there were more than 100 brands of ketchup sold in the United States), and baked beans; others focused on niche markets such as specialty pasta entrées and seafood dishes.

Some, such as H. J. Heinz, began with a limited product line and expanded as rapidly as possible across numerous product lines. Others, such as the Campbell Soup Company, began as general canners and reduced their lines to a few specialized products in which they excelled.

The Campbell Soup story is a classic example of a family of entrepreneurs developing a market for an item whose consumption was declining because it took too long to prepare. The company had its beginnings when Abraham Anderson, a Philadelphia tinsmith who had been manufacturing iceboxes, decided to cross the Delaware River and go into canning poultry in Camden, New Jersey, in 1862 (Collins, 1994). Anderson invested $400 in his infant company. He soon acquired a partner, Joseph Campbell, and the new company, Anderson & Campbell, increasingly concentrated on vegetable processing. Campbell bought out Anderson in 1871 and formed the Joseph Campbell Preserve Company, which continued to grow in both total sales and product lines. Mincemeat and Beefsteak (tomato) Ketchup were its two most important products in the 1890s. Arthur Dorrance became vice president and treasurer in 1891 and took control of operations at Joseph Campbell's retirement.

The future of the company changed forever in 1895 when John T. Dorrance, Arthur's nephew, stopped by the factory one day while sailing on the Delaware River and discovered that the company did not make soups. Graduating with a Ph.D. from the University of Göttingen (Germany) in chemistry, he ignored his family's hopes that he go into college teaching and accepted a $7.50-a-week job as a chemist with the cannery. His position was so lowly that he was required to supply his own equipment, but he was given a percentage of the profits of all of the new products he developed. Less than twenty years later he was sole owner of the Joseph Campbell Preserve Company and business was booming.

John T. Dorrance created five condensed-soup flavors in his first year with the company: tomato, vegetable, chicken, consommé, and oxtail. His strongly flavored, simple soups blended "English thoroughness and French art," in the words of Fannie Farmer. The presence of the seemingly out-of-place oxtail, long a company favorite, on the list stems not from national but from local popularity. The company soon developed twenty-one soups and was selling more than 16 million cans of condensed soup by 1904. Although new soups were developed, the number was limited to twenty-one for almost a century. Soup became such an integral part of the company's self-image that the name was changed to the Campbell Soup Company in 1921. The Dorrances also reduced the number of products the company produced until only soup and pork and beans remained. Pork and beans were cooked and canned on Mondays while the soup kitchens were preparing soup stock for the other four days of the week.

John Dorrance was convinced that with a little encouragement Americans would add a soup course to their evening meal. Much of the company's advertising was targeted at expanding the original market for the product—primarily a luncheon food for children in winter. Advertisements exhorted housewives with "I couldn't keep house without Campbell's Tomato Soup" or "It seems to fit exactly into every kind of menu. And it makes the whole meal taste better and go better." The company's first cookbook, *Helps for the Hostess,* was published in 1916, but the real advertising success was the introduction of the Campbell Kids. They symbolized the core market, which has continued despite the company's efforts to add a soup course to the dinner menu.

The use of condensed soup as a sauce was also introduced in the company's first cookbook, though the company resisted exploiting the success of this concept in favor of concentrating on soup sales. Things changed after Arthur Dorrance's death. The company formed a home economics department to develop new recipes and products as well as produce cookbooks to encourage the use of Campbell's products. No small business, the cookbook department sold more than a million books per year during the 1950s. The company also began developing more creative, nonsoup ways to utilize condensed soups in America's kitchens.

The publication of *Cooking with Condensed Soup* in 1952 had an amazing impact on American cooking. It promoted such favorites of the era as Heavenly Ham Loaf, Green Bean Bake (green beans, cream of mushroom soup, and a can of french-fried potatoes), and the all-time favorite, Perfect Tuna Casserole. Sales of condensed soup increased to an estimated 1 million cans per day just for use as a sauce in creating these new family favorites. For many years, it was almost impossible to visit friends for dinner without encountering one of these Campbell's taste treats. Food writer Jo Brans wrote of an increasing fear among children of visiting friends for dinner and being forced to eat "funny food, like lasagna or tuna fish casserole, or some strange vegetable" (Collins, 1994).

Despite its success, Campbell Soup was never able to convince Americans to add a soup course to their evening meal. The company spent millions of dollars for advertisements, training, production of cookbooks, menu planners, and a host of other promotions in the quest. The American public may have been captivated by the Campbell Kids and "M'M! M'M! Good!" but homemakers continued to serve soup to kids only for lunch and snacks. The American democratic belief that meals were single-course events was unas-

Perfect Tuna Casserole

1 can (10 1/2 ounces) condensed Cream of Celery or Mushroom Soup
1/2 cup milk
1 can (7 ounces) tuna, drained and flaked
1 cup cooked peas or green beans
1 1/4 cups slightly crumbled potato chips

Blend soup and milk; stir in tuna, peas, and 1 cup potato chips. Spoon into a 1-quart casserole. Sprinkle top with remaining potato chips. If desired, use whole chips on top of casserole rather than crumbled. Bake in a moderate oven (375° F.) about 25 minutes. Makes 3 or 4 servings.
(Campbell's Soup Company, 1952)

sailable. In a society where even coffee was served as a part of the main course, a separate soup course never had a chance.

The Evolution of a Prepared-Food Conglomerate

The period after World War II was characterized by the consolidation of a wide variety of related and unrelated companies under a single corporate umbrella. One of the best examples of these conglomerates that successfully utilized synergy to become a dominant force in shaping the food industry was the Beatrice Company. Some of these food companies have been spectacularly successful, most notably the Philip Morris Companies and ConAgra; others were eventually either swallowed up by even larger organizations such as the Pillsbury Company and Kraft or just self-destructed; Beatrice is an example of the latter. The process that took place during this period created a new approach to the creation of food products and their marketing that still shapes our diets today. Thus it is worthwhile to look at one of these companies to see how it evolved.

The Beatrice Company was founded in 1894 when Thomas Haskell formed a partnership with William Bosworth to distribute poultry, eggs, butter, and produce from farms around Beatrice, Nebraska. The distribution of farm products continued to remain at the core of the company's business, although it built a creamery in Beatrice to better control the quality of its butter, which was being marketed throughout the Midwest. An example of the company's attitude toward its producers is well illustrated in its solution to the spoilage problem of the skim milk that was returned to the farmers for use as livestock feed. Rather than ignore the problem or require the farmers to purchase separators at inflated prices, Beatrice began a program of financing the sale of more than 50,000 separators to their producers at cost.

Dairy-product processing, distribution, and the ownership and management of public cold-storage warehouses remained the core of the company's business until 1955 with the exception of La Choy Food Products. The era of the conglomerate led the Beatrice managers to set the company on an acquisition spree of dairies and other food businesses. The list of subsidiaries is lengthy and includes Krispy Kreme doughnuts of Winston-Salem, North Carolina, and the Red Wing television production company, created by comedian Garry Moore. Three company policies guided the operations of these new acquisitions: Leave the original management in place with little external control; encourage these typically regional companies to expand their markets to national levels; and provide the cash, marketing savvy, and clout to assist these subsidiaries in creating a national presence. The beneficiaries of these policies were many, most notably Eckrich, Tropicana, and Shedd-Bartush Foods. Eckrich, a regional maker of hot dogs, kolbasa, and other ethnic sausages and luncheon meats was able to expand distribution nationally, even to places that had never heard of bratwurst or kolbasa. Tropicana significantly lowered costs with Beatrice's network of cold-storage warehouses and warehouse commitments and became the first national distributor of "not from concentrate" citrus juices. Shedd's Spread was able to use the new marketing expertise to become one of the nation's best known oleomargarines.

La Choy, one of Beatrice's earliest acquisitions, is an especially good example of the role that large prepared-food companies could play in helping smaller companies create markets for little-known products. La Choy was founded in Detroit, Michigan, in 1920, just as anti-Asiatic sentiments in the United States were reaching a fever pitch. The company faced incredible hurdles. Founded by a European and a Korean, La Choy Food Products attempted to market a product to a population that had little interest in the consumption of Chinese foods in restaurants and almost no interest in doing so at home.

La Choy Food Products was reasonably successful despite these hurdles, focusing first on producing and selling bean sprouts, then adding soy sauce, and finally expanding to an entire line of Chinese foods. The company grew to the point that it could build its own Detroit processing plant in 1937, and its future seemed bright. But four years later the federal government classified the production of Chinese food as nonessential to the war effort. Unable to secure tinplate for cans, La Choy sold the plant to the government, which converted it to a munitions plant.

Disheartened, the managers at La Choy continued producing noncanned products until approached by Beatrice in 1943. Recognizing La Choy's quality management team and long-term potential, Beatrice helped the company find and purchase an abandoned steel mill in Archbold, Ohio, which was revamped into a vegetable cannery. The new plant was used to process canned tomatoes and Vegemato juice for the remainder of the war, during which time the company gained additional canning expertise. The acquisition of a new marketing manager with strong grocery-chain experience recommended by Beatrice made all the difference, and production and sales of Chinese specialties exploded at war's end. By 1985 La Choy held 40 percent of the specialty canned-entrée market in the United States and about 25 percent of the specialty frozen-entrée market. The company's sales passed the $135 million mark during that year.

The La Choy experience demonstrates the positive side of the conglomerate period in the food industry. Large companies were able to focus their expertise and financial resources on smaller companies and jump-start their profitability. All too often these large corporations understood little about the food industry, however, and attempted to use their chemical or heavy-industry management and sales experience in managing food companies. Many were spectacular failures, but recognition of the advantages of related product lines continued the creation of ever larger food producers. The complex interrelationships that have been created by the evolution of companies like Philip Morris and ConAgra are so extensive that it is difficult for smaller competitors to remain viable.

The Institutional and Restaurant Food Supply

A large share of the prepared-food market has always been in specialty foods for the institutional and restaurant trade. Indeed, many of both the earliest canned and frozen products, such as lobsters, oysters, and salmon, were largely destined for restaurants and institutional food service.

Restaurants need to have guaranteed sources of food supplies before items can be placed on menus, which are often used for years at a time. Partially or totally processed foods that were canned or later frozen allowed them the freedom to make long-term decisions about the prices and character of the foods they could offer.

The rise of chain restaurants with hundreds or thousands of units, all serving exactly the same menu items, created a further need for guaranteed sources. It is not uncommon for a large restaurant chain to serve millions of pounds of specific items when they are placed on special. Even prosaic foods like bacon can become in short supply when a chain like McDonald's or Wendy's makes bacon cheeseburgers a promotional special. Although the largest chains now attempt to protect themselves from price increases by purchasing their supplies prior to the promotion, they are not always able to estimate the impact of a marketing campaign. Apple South, the largest franchise of Applebee's restaurants, did a television promotion on a riblet special several years ago and prudently purchased more than 5 million pounds of product prior to the promotion campaign. It sold its entire supply in the first half of the promotion period. The price of the additional product had already been driven upward by the initial purchase, and the company was forced to abandon the too-successful campaign weeks earlier than intended.

A variety of specialty foods have been developed specifically for these markets. The current french fry came into existence in the early 1960s when McDonald's began expanding so rapidly that the company could not purchase sufficient supplies of high-quality potatoes late in the season before the new crop was available. Ray Kroc, then only the national franchisee for McDonald's (there were a few stores operated by others prior to his purchase of the national franchise), approached J. R. Simplot of Boise, Idaho, the inventor of frozen french fries. Kroc needed a french fry that would meet the high standards of the McDonald brothers, who considered consistent high-quality french fries to be one of their most important marketing tools. Simplot spent more than $1 million of his own money creating a frozen, blanched product that when cooked was actually better than that made from fresh potatoes. It was so good that it became the industry standard, and the fresh-potato french fry has virtually disappeared from commercial use. Simplot, not surprisingly, is a billionaire today. His privately held company is the world's largest supplier of frozen potato products and one of the largest suppliers of frozen vegetables. Few Americans have ever seen or heard of the Simplot company, which has sold only to the institutional trade until recently.

The high cost and short supply of trained kitchen help have also made prepared foods more important in the restaurant kitchen. Trained chefs demand and receive high wages for their imaginative work. An increasing percentage of the food items have been simplified so that much of the work involves little more than heating the food and dressing the plate. More and more of the food passing through the kitchen is partially or fully prepared off site. No longer do armies of sou-chefs spend their days chopping lettuce, peeling potatoes, and making sauces as they would have in the past. In all but the most sophisticated restaurants today, the salad enters the door pre-chopped and mixed with garnishes, the sauces arrive prepared in frozen tubs or large cans, and increasingly even comparatively sophisticated entrées arrive flash frozen. Finally, the remaining workload is subdivided in such a way that few who work in the kitchen need know anything about the cooking process except when to turn the machine on or off or turn the meat.

A secondary element in the rising use of outside prepared food is the chain restaurant's need for consistency. Whether the food is prepared in the chain's commissary or contracted to a third party, the use of prepared foods means that the restaurant operator or franchiser is guaranteed that every plate of spaghetti in every restaurant from Orlando to Seattle is going to taste exactly the same. The franchiser is also assured that every ingredient in that spaghetti sauce meets company standards. The unit operator simultaneously is comforted by the knowledge that when a member of his kitchen help does not show up for work, his job can easily be switched to another low-paid worker. The owner will also not have to spend hours each day overseeing the creating of sauces and other meal elements that take hours to prepare; nor will he need to worry about price fluctuations, because he has contracted for the delivery of his raw materials at a predetermined price.

Harvey Houses were the first chain to utilize central purchasing of foodstuffs and to oversee delivery to all of its units over a wide geographic area; Howard Johnson's restaurants were one of the first large chains to utilize the commissary approach to controlling costs and quality. Howard Johnson's adopted a central commissary system during the 1920s with everything except the coleslaw being prepared off site. The rise of quick-service restaurants broadened the concept. Col. Sanders initially sold his secret seasonings only to restaurants but ultimately realized that the only way to maintain quality was to provide all the principal ingredients. Today, custom distributors contract with quick-service stores to coordinate deliveries from warehouse to storage areas on a just-in-time basis.

Most prepared-food manufacturers produce large lines of institutional packs, many of which are not available to the general public in retail stores. Institutional packed foods are not composed of lower-grade fruits, vegetables, and entrées that the packers feel they cannot sell on the retail market; rather they are just packed in quantities convenient for volume operations. Bags of fifty frozen beef patties or breaded chicken breasts, trays of a dozen or more frozen croissants and muffins, or boxes of hundreds of dough balls

Lunching Across North Dakota

It was well past noon and John and I had been driving across central North Dakota for more than two hours without sighting a town or restaurant. The arrival in Pillsbury wasn't entirely by accident, of course; how could I resist traveling at some time in my life to a town named after my family—even if it was in east-central North Dakota? And here we were at last. The town was little different than the other dying towns we had been passing over the past several days. Pulling past the grain elevators, we decided to wait for the requisite pictures in front of the sign until after we had found something to eat. We drove through town looking for likely places but saw only boarded-up houses, boarded-up stores, a post office–general store, and a grungy-looking bar.

Pulling into the post office, general store, and whatever, we wandered in and checked out the shelves. The owner probably thought we were going to rob the place, never dreaming we were just two writers looking at the American landscape. We saw pretty much what we expected and had been seeing since we hit the American grasslands a couple of days before, but wait. What was that in the frozen-food case? Burritos? Tortillas? Frozen chili rellenos?

Trying to look casual, I wandered over to the owner at the cash register: "Looks pretty quiet here."

"Yeah, it's pretty quiet. The elevator is about all there is anymore."

"Anyplace to eat here in town?"

(continues)

that will become oven-fresh yeast rolls served piping hot to the customers from the restaurant's own ovens are typical of those products that dominate the institutional food industry.

It was generally believed that the public would not purchase these items because of the size of the packs and concern about the quality of the unknown products. Recent experience with warehouse grocery and wholesale buying chains, however, has discredited this belief. Consumers have discovered that many of these items are better than those they can prepare themselves. An extension of this concept has led to the rise of gourmet "meal replacement" stores, which specialize in high-quality, piping-hot entrées and side dishes targeted at customers who are tired of eating in restaurants but still don't have time to cook at home.

These products have become increasingly attractive to busy households with little time to purchase, assemble, and cook individual items. Small bags

(continued)

"Well there's Rock 'n Rodney's over there. It's pretty good."

"We saw that; we weren't sure it was open or that we would want to eat anything they had."

"Oh, don't let looks scare you. They got great food."

"Well maybe we'll give it a try. By the way (sneaking in the real question at last), I see you have a box of frozen burritos over there. Many Mexicans move in here to work the wheat?"

"No, we don't get any of those fellows up here. I eat the burritos myself. In fact, I'm going to have burritos tonight. Probably sell a case a week of those things."

"Burritos are kind of a strange thing way up here."

"Well, I guess so," he said, pausing to think a moment, "but you know it's kind of nice to get a little variety now and again. We don't get much exotic food up here. If you want something different you got to do it yourself."

I thanked him, purchased a package of cheese crackers (I didn't want it to appear that I had really come in to pry), and reluctantly moseyed over to Rodney's.

Now, let me say that if you are ever in Pillsbury, North Dakota, skip the frozen burritos and the cheese crackers and go directly to Rodney's. The ambiance is basic, but the hamburgers are world class. They also sell garish T-shirts with an Elvis-looking Rodney telling everyone to come on down to uptown Pillsbury.

of salad mix and peeled carrots and other vegetables have become an integral part of every supermarket produce department. Increasingly, middle- and upper-middle-class dual-worker households pay premium prices for prepared foods to reduce the money (and, more, important, time) spent on dining away from home.

Busy households have also meant that increasing numbers of Americans know little more than the rudiments of cooking. Forty years ago the daughter of the family (or youngest son, in my case) often spent hours in the kitchen with her mother learning the mysteries of how to cook all of the family favorites. The arrival of Little League and then girls' sporting teams meant that less and less time was available to spend in the kitchen learning cooking techniques. All of the blame cannot be placed on after-school activities, however, as Mom probably wasn't there slaving away over the stove anyway; rather she was at work. The result has been a very significant decline in cooking knowledge in the typical household. Coupled with the attitude that cooking is boring, this has led to ever greater demands for prepared foods.

Ultimately, however, prepared foods did not transform the American diet so much as the transformation of the American household demanded the creation of an entirely new set of foods. Prepared foods may have begun to reduce the housewife's workload, but ultimately they provided a window on foods that had never been available to the mass American population. Exotic entrées became available in the smallest community in America, though clearly the concept of exotic must be tempered. No one of Chinese descent would ever mistake La Choy chow mein for a product of the homeland; nor would any Italian recognize Franco-American Spaghetti-O's as Italian in origin—each has been heavily Americanized, or actually created for, the mass American market. Yet the introduction of each of these foods has encouraged those Americans unfamiliar with and fearful of "foreign" cuisines to be a little more adventuresome. The availability of cream of mushroom soup for making tuna noodle casserole, Dinty Moore beef stew, smoked oysters in a can, and frozen burritos has reshaped the way we eat.

5

Marketing to the Masses

> *During the past year the [food distribution] trade has been well aware of the fact that many industries are long on production and short on distribution. The economic distribution of farm products is today the world's worst problem. . . . For years the Department of Agriculture has been dumping tons of literature on the farmer, telling him how to produce, while the farmer has been dumping tons of products in the nation's garbage can for want of a market.*
>
> —***The Chicago Packer*, 1915**

Few Americans have ever been totally self-sufficient in producing their own foodstuffs. In fact, average family food purchases exceeded the value of those produced at home when the government first tabulated this statistic for the census of 1870. Obviously the government had no realistic way to accurately estimate the real market value of food produced at home for personal consumption, but these imaginary numbers at least provide a baseline for comparison. The amounts of food purchased and produced at home remained comparatively equal until just before World War I. Home food production remained a strong component in the dietary regimes of millions of Americans in the years thereafter, but the final migration to the city, the shrinking in size of town building lots, the increasing specialization of farming after World War II, and myriad other forces increasingly brought an end to home food production for most Americans. Today purchased foods represent almost 98 percent of all food consumed in America.

Obviously the food-distribution system—getting the food from farm to processor to table—plays an important role today in shaping the price and quality of what Americans consume. There are hundreds of wholesale distributors today, but the vast majority of all food arriving on the loading docks of most grocery stores is taken there by trucks owned and operated either by the company that owns the grocery store, a commercial grocery wholesaler who has a long-term relationship with the store (which may also bear its name), or a grocery distribution cooperative that acts as an agent for a group of independent stores. Kroger, Safeway, Publix, and Albertson's are grocery chains with their own distribution network owned by the company. Super Valu and Piggly Wiggly are examples of chains in which the stores are independent but have a semipermanent relationship with the grocery chain. IGA is probably the best known of the independent grocery store groups that belong to cooperative purchasing associations. Because of this phenomenon, the marketing discussion will open with an examination of the role of the retail store in shaping the American diet will be followed by an examination of advertising in the food and kindred products industry.

The Retail System

Manifestly the most wasteful part of our economic system is that concerned with distribution. . . . Every investigation of the high retail price of agricultural produce—meats, cereals, dairy, vegetables, or fruit products—has shown an enormous gap between the price received by the first producer and that paid by the ultimate consumer.

—***Report of the Massachusetts Commission on the Cost of Living*, 1910**

There were 101,285 retail grocery stores doing $402 billion of business in the United States in 1994. The retail sale of food is big business, and it could be argued that its very structure shapes what we eat. In the beginning there were no stores at all; the transfer of excess foodstuffs from producers to consumers was handled either through importers who sold in bulk or at weekly markets in the town square. Other foods entered the system as farmers traded excess food for goods and services with neighbors.

The evolution of a system of retail sale of goods by professional agents rather than the people who produced them was a slow process. There were few freestanding, general merchandise stores until the eighteenth century, and these primarily sold farm equipment, hardware, utensils, and dry goods. Imported foods were available, but most locally produced goods continued to enter the system through direct trades and sales between purchasers and producers.

The general merchandise store primarily provided goods not available to the community, and typically most transactions were on credit. Local farmers were extended credit throughout the year, settling their accounts at harvest time. This seemingly generous arrangement was actually to the advantage of the merchant: Interest rates were low, specie was in short supply, the retailer probably was obtaining his goods on credit from his suppliers, and there was an implicit assumption that a farmer would both sell his harvest through and concentrate his retail purchases in the place that extended him credit.

The increasing importance of purchased food allowed the evolution of freestanding grocery stores during the nineteenth century. Early grocers tended to handle only nonperishables, an extension of their early general store activities. Meat was sold by the local slaughterhouse, which eventually opened a separate retail store; fresh produce continued to be traded on the square or purchased from local farmers known to produce excess quantities of these goods. Even in urban areas produce was sold from carts and stands by individuals who obtained the fresh goods from farmers. Most baked goods were also sold directly by the bakers.

The creation of an extensive system of wage labor with the Industrial Revolution brought increasing numbers of consumers to retail food outlets with cash to purchase their groceries. This allowed consumers to begin discriminating between retailers in terms of price, quality, and selection on even individual items. The Great Atlantic and Pacific Tea Company was one of the first successful grocery chains; its development well illustrates the early evolution of the grocery store as we know it.

George Gilman, a New York City coffee and tea peddler, began to take advantage of the new retail environment in 1859 by offering tea and coffee from his cart in various parts of the city at a discounted price. Expanding business allowed him to purchase more wagons and eventually open fixed-site stores. He was operating more than 100 stores and uncounted peddler wagons when his junior partner, George Huntington Hartford, took over the business in 1878. Capitalizing on annual gross sales exceeding $1 million, Hartford expanded into a full line of dry groceries to become one of the largest grocery store chains in the nation. The Great Atlantic and Pacific Tea Company, the A&P, was incorporated in 1901 with about 200 stores and sales of $5.6 million.

All grocery stores operated much the same in those days. Customers entered the store and presented the storekeeper with a list of goods desired. Brand names were virtually nonexistent until the 1880s, and most goods were purchased in bulk by the grocer, who then sold them in smaller units. The clerk took the order and sent a boy into the warehouse or to the back of the store to fill the order. The customer could not browse or even see the condition in which most food was kept. Many prices were set by the manufacturers and competitive pricing was inconceivable.

The cash-and-carry concept first appeared soon after 1900. Relieved of absorbing credit and delivery costs, these new stores could lower retail prices because of lower labor costs and a faster return on their capital. The Great Atlantic and Pacific Tea Company was one of the first large-scale users of the new system. Hartford's sons convinced their father to test market a small cash-and-carry store in Jersey City, New Jersey, in 1913. The store was a success, and they soon convinced him to open 7,500 of these neighborhood stores over the following three years. The plan was simple. The company would obtain short-term leases in likely neighborhoods, investing no more

The invention of the self-service grocery in 1916 played an important role in expanding the power of brand-name foods. (Piggly Wiggly Stores)

than $1,000 in each location. Low prices and bright red storefronts brought instant attention, and hiring only a manager and an assistant kept overhead low. Poor sales led to immediate closings with a minimal loss. A&P profits went from just under $2 million in 1915 to $8.4 million in 1921.

Cream of Wheat initially was so incensed at the A&P policy of flagrantly ignoring its pricing policies that it initially refused to sell additional product to the company. This did not last long. A&P sales of Cream of Wheat's competitors' products expanded so rapidly that Cream of Wheat soon realized that it had to accept the situation or go out of business; A&P's competitors were adopting the same strategy as well.

The self-service grocery store had an even greater impact on food retailing. The concept was patented by Clarence Saunders of Memphis, Tennessee, in 1916. Customers were allowed to wander about the grocery for the first time to select their own items. Saunders had opened 125 of his Piggly Wiggly stores by 1919 and began licensing the concept to others. A store-fixtures factory was built, and equipment to open an entire self-service cash-and-carry Piggly Wiggly could be loaded onto a single boxcar and delivered to the purchaser's town in a few weeks. Saunders may have had a patent, but the basic simplicity of the concept meant that it was only a short time before the entire industry began switching to this low-cost way of doing business, especially once the potential effect of impulse purchasing on sales was recognized.

The self-service concept revolutionized food marketing. Thomas Crowell and some other manufacturers had given some thought to packaging prior to this time, but most had not. The consumer's new freedom to walk down an aisle and select products on the basis of price or package advertising also increased public awareness that competitors existed; thus competitiveness increased as well. Consumer awareness also increased the potential of new products to be successful in the marketplace, the potential power of brand identity, and the variety of foods that were purchased on impulse. The manufacturer could appeal directly to the consumer through its packaging rather than being dependent on the retailer to give demonstrations or post advertising literature.

Automobiles began appearing in large numbers during the 1920s but did not become an important force in the food industry until after World War II. Food stores maintained traditional retail business hours, and if a family was affluent enough to have an automobile, it was used to transport the family breadwinner to and from work during the week. Trips to the A&P or Safeway had to wait until Saturday morning, when the family piled into the car for a trip downtown to purchase groceries, do a little general shopping, and possibly stop off at the soda shop for a cherry cola. The remainder of the family's food purchases during the week were made at small neighborhood establishments, often by the children on their bicycles. These small neighborhood stores carried only basic, fast-moving items that were popular in their specific trade areas. Variety was slow in coming.

The arrival of a second automobile allowed the housewife to travel longer distances to larger stores with greater variety and presumably more competitive prices. These new stores, often at the edges of the business districts, were cleaner, brighter, and offered many foods that were not available elsewhere. These supermarkets had their foundations in Clarence Saunders's Jefferson Street Piggly Wiggly but actually date from 1932, when Michael Cullen opened his first King Kullen store in Jamaica, New York. His gigantic store, about 10,000 square feet, was about ten times larger than his typical competitor of the day. It included a fresh-meat section as well as the traditional groceries. There were 1,200 of these new large stores in 84 cities around the nation by 1936. They were important enough to receive their own nomenclature the following year—supermarkets.

The supermarket had an immediate impact on the grocery industry by raising the entry cost of competition. Small grocers were shoved into less and less competitive locations. They were also saddled with higher inventory costs. Manufacturers typically gave volume discounts to large purchasers, allowing the low-overhead, high-volume stores to have yet another cost advantage over their smaller brethren. The problem was so blatant that Congress began an investigation of the rebate and unfair marketing practices of the A&P chain during this period. It was discovered that A&P had agreements with at least 114 of the 170 major food manufacturers in the country, although about 60 percent of those were available to other large retailers as well. Though basically an antitrust action, the case is most interesting today as a milepost in the changing character of business practices. A&P did indeed receive preferential treatment from some of the nation's largest food producers but often because they were the single largest customer for many of these companies. Further, A&P used its volume purchasing power to acquire large quantities of manufacturers' excess goods at highly favorable prices. Using the lower purchase price to allow lower sale prices, the chain could then sell the products as apparent loss-leader sale items. A&P virtually rescued some processors from their own incompetence by relieving them of their excess inventory. The downsides were that A&P could potentially use its lower costs to force competing retailers out of business and pressure processors to make decisions in its favor. And in practice, the company did go into production itself if it could not find processors willing to meet its price points; generally, however, processors quickly realized the precarious nature of their position when dealing with these large retailers. Interestingly, A&P was able to create such action and reaction without ever exceeding a 10 percent national market share.

The supermarket concept did not hit its stride until the postwar period. Ten thousand supermarkets were in operation by 1950. The supermarket did not totally force smaller units out of business as predicted; rather it led to a complete restructuring of the retail grocery concept. The *Directory of Supermarket, Grocery & Convenience Store Chains* currently recognizes six classes of units: superstores, supermarkets, convenience stores, warehouse stores, combo stores, and gourmet stores. Supermarkets, with 17,370 units in 1995, continue to dominate total sales.

The new superstore category, made up of stores exceeding 30,000 square feet and sales of $10 million per year, are now appearing in larger cities. Whereas most of the 5,271 superstores are little more than overgrown supermarkets, increasing numbers are oriented toward new combinations of activities. Harry's Farmers Markets is a classic example of a specialized superstore. Harry's primary business is fresh produce. The flagship store devotes about one-half of its 120,000 square feet of space to fresh fruits, vegetables, and other produce. The stores typically carry five or six types of lettuce, as many as fifteen types of mushrooms, and twenty kinds of peppers at any one time. The fish department typically sells forty or more types of fish including live lobsters, Dungeness crabs, crawfish, mussels, four types of clams, and oysters. An assortment of cookbooks and cooking tools are carried, but unlike in most superstores, nonfood items amount to only about 17 percent of the chain's approximately $117 million in annual sales (*Directory of Supermarket, Grocery & Convenience Store Chains* [hereafter *Directory*], 1996). The emphasis is on specialty exotic foods rarely available in competing stores. As a result, classic dry grocery brands and items are few and far between. The store does carry ketchup, mayonnaise, and Pepsi products but is more noted for thirty kinds of barbecue sauce and eight brands of jerk seasoning. Harry's operates only three of these stores in Atlanta as well as a handful of gourmet miniunits (Harry's in a Hurry) strategically placed in upscale neighborhoods. The Boston Chicken Corporation, however, has recently purchased rights to build additional Harry's stores outside of Georgia and Alabama.

The convenience store, a reinvention of the neighborhood store, usually offers consumers a combination of groceries, gasoline, and ready-to-eat food (microwavable, cold, or increasingly cooked to order). It is numerically the most common store form with 68,624 units in 1995 but has a lower sales volume than the larger stores. The C-store, as it is increasingly called, has grown primarily because gasoline companies perceived that bundling the two low-labor services could take place with minimal increase in cost. Gasoline is generally sold at very competitive prices. Customers purchase enormous quantities of soft drinks, fast foods, magazines, and sundries while completing the gas purchase. More than 50,000 units sell prepared food, books and magazines, and automobile supplies. Other popular amenities include automatic teller machines (as isolated, outdoor, drive-up windows are increasingly perceived as unsafe), video rentals, and in-store bakeries. Incredibly, 167 have floral and horticultural departments and 1,765 have in-store restaurants, usually a fast-food chain (*Directory*, 1996).

The grocery business is highly diffuse with more than 90,000 stores servicing this retail sector; yet it is highly concentrated in that even the 200th largest chain grosses more than $170 million per year (see Table 5.1). The number of units in a chain can be deceptive because of the wide range of sales volumes even within store types. For example, the smallest of the top 200 supermarket chains, biggs of Milford, Ohio, has only seven units but grossed a little over $0.5 billion in 1995. In contrast, units and sales are closely aligned in the convenience-store business. The Southland Corporation, known as 7-Eleven in most areas, had 5,796 stores with grocery sales of $6.7 billion in

TABLE 5.1 Leading Supermarket and Grocery Chains, 1995

	Sales ($M)	*Units*
The Kroger Co., Cincinnati, OH	22,384	2,225
Safeway Inc., Oakland, CA	15,215	1,075
American Stores, Salt Lake City, UT[a]	14,476	911
Albertson's Inc., Boise, ID	11,284	676
Winn-Dixie Stores, Inc., Jacksonville, FL	11,082	1,161
Great A&P Tea Co., Montvale, NJ	10,480	1,135
Food Lion Inc., Salisbury, NC	7,610	1,028
Publix Super Markets, Inc., Lakeland, FL	7,473	445
Vons Companies, Arcadia, CA	5,075	388
H. E. Butt Grocery Co., San Antonio, TX	4,450	347

[a]Holding company that operates 433 Lucky Stores (California), 247 Acme Markets (Middle Atlantic), 198 Jewell and Jewel Osco Stores (greater Chicago and Southwest), and 33 Star Markets (Massachusetts). Safeway has subsequently purchased Vons.

SOURCE: *Directory of Supermarket, Grocery & Convenience Store Chains*, 1996.

1995. Circle K, the second largest, has 2,508 stores and grossed about half as much in sales, $3.2 billion, in the same year (*Directory*, 1996).

The behemoths dominate the industry despite the large number of competitors. The top ten supermarket chains account for about one-third of sales and the twenty-five largest companies account for 46 percent of all sales. The very size of these chains allows them to build strong relationships with suppliers. Simultaneously their very size encourages them to reduce their number of vendors; to cater to the largest manufacturers, who potentially could give them preferential treatment; and to ignore small producers, who could not possibly supply sufficient goods to service all of their stores. The Kroger Company, for example, was very slow to add free-range chicken, organic beef, or oriental vegetables to its stores but has been very quick to add precut meat, home meal-replacement operations, and Tyson's tortillas. It easily caters to the mainstream but has trouble servicing the counterculture. The major companies typically carry few products from regional canners and processors and even fewer regional specialties unless the store or departmental manager is an aficionado of the items in question. This is not to suggest some dark supermarket plot; rather it is an illustration of the economics of a $400 billion business. One of my local Kroger stores, for example, carries a barbecue sauce manufactured in the kitchen of someone who lives down the street from the store. But that kitchen belongs to a regular customer of that store, the product is unusually good, and that store manager is atypical.

Most stocking decisions by large retailers today are made by a set of computer programs at company headquarters that automatically restock when supplies get low and stop carrying items when sales decline below preset square-footage sales numbers. Shelves of company policy manuals typically determine what is to be carried, how much space is to be devoted to what, and the brands and package sizes to be stocked. The specialty shopper quickly learns that his Lucky Store does not carry that desired off-beat item and shops for it somewhere else. This creates an automatic catch-22 situation wherein the store does not realize that any demand at all exists so is even less likely to carry the item, whereas specialty ethnic and gourmet stores, more tuned to the local market, are presented with an opportunity to meet the demand.

Advertising

More than $425 billion was spent on advertising in 1995, and a large part of that was spent in trying to change Americans' minds about what they were going to eat. Four of the top ten and eight of the top twenty advertisers are in the food business. Further, the rate of advertising has been increasing at a phenomenal rate as the consolidation of companies increases competition (see Table 5.2). My daughter could spot a McDonald's golden arches before she could say "Dada" and by age three could spot one before me. Even a cursory examination of the growth of the food industry over the past 100 years demonstrates that marketing strategies, especially advertising, have played a crucial role in what we eat and where we eat it.

Store signs, the first advertisements, had appeared by the mid-seventeenth century and were followed by many other types of other signage. Early printed advertisements tended to concentrate on shipping schedules and notices about lands for sale and runaway slaves. Paper was scarce after the Revolution because of a shortage of rags as raw materials, and most newspapers refused to run anything more than one-inch advertisements with the same type sizes as regular news stories. Broadsides, flyers that could be posted on trees, posts, and buildings, became the most common way of advertising, but even these advertisements primarily concentrated on making the reader aware that goods were available at a particular location. A frequent addition to almost all urban advertisements was "We will sell low for CASH, or Country Produce," suggesting that nearby farmers were an important source of fresh produce. Retail prices were often set by the manufacturers, and discounting as such was unknown.

TABLE 5.2 Leading Food Advertisers, 1995

Rank	*Expenditures ($B)*[a]	*Select Food Products*
2. Proctor and Gamble	1,501	Folgers, Crisco, Pringles, Citrus Hill
3. Philip Morris	1,400	Kraft, Post, Miller Brewing, Tombstone Pizza
5. PepsiCo	707	Pepsi, KFC, Frito-Lay, Pizza Hut
10. General Mills	526	Betty Crocker, Gorton, Big G Cereals
13. Kellogg	494	Cereals, Eggo Waffles, Pop Tarts
17. Nestle, S.A.	476	Maxwell House, Nestle, Carnation, Stouffer
18. McDonald's	447	Quick-service restaurants
19. Unilever, PLC	432	Lipton's
20. Grand Metropolitan PLC	364	Pillsbury, Burger King, Green Giant, Pet
29. American Home Prod.	323	Chef BoyArDee, Gulden's, Jiffy Popcorn
30. Anheuser Busch Cos., Inc.	317	Budweiser, Eagle, Rainbow bread
35. RJR Nabisco	282	Nabisco, Life Savers, Grey Poupon

[a]Reflects total corporate expenditures, not just funds spent advertising food items. Consumables are not the primary business of Proctor and Gamble, Philip Morris, Unilever, and RJR Nabisco.

SOURCE: Gathered by the author from multiple sources.

The big change in marketing after midcentury was the creation of recognizable brand names. The shift from handicraft to factory production meant that goods were shipped increasing distances; thus the consumer was less likely to have personal knowledge of the producer. Most food goods were shipped in bulk—barrels of flour, oatmeal, salt pork—and the consumer asked for a slab of bacon, not for a slab of Hormel bacon. Many food products acquired names based on their region of origin—Smithfield hams, for example—but it soon became obvious to manufacturers from Cyrus McCormick to Ferdinand Schumacher that if they could attach a brand name to their product it would increase sales. Schumacher placed what is believed to be the first advertisement for breakfast cereal in the *Akron Beacon* in 1870. The advertisements apparently were successful, and sales of his Avena oatmeal increased rapidly. Schumacher may have been known as the king of oatmeal in his day, but his successor, Henry Crowell, was the man who shaped cereal advertising, and to a large degree all food advertising, for the remainder of the century

Oatmeal was a commodity sold by the barrel when Crowell entered the oatmeal business. Schumacher and other cereal manufacturers had had little control over the condition of their product when it was sold to the consumer. Schumacher sold a small amount in two-pound glass jars, but most was sold to wholesalers and retailers in the fall and sat in the barrel until sold. Insects, rodents, and dirt all mingled with the product. Crowell's strategy was to distinguish his product by selling it in individual sealed packages and to guarantee purity. He selected a cylindrical box that would hold two pounds of product and was printed with a picture of a Quaker holding a bowl or box of oatmeal in one hand and a scroll in the other on which the only readable word was PURE. He used a dark blue, red, and yellow to make the package stand out on the shelf even though self-service stores were still a half-century in the future.

Crowell completed the campaign with saturation advertising. He painted signs on barns; he ran special trains festooned with signs and filled with oatmeal; he invented the concept of a premium (glass, bowl, saucer) in the box; he festooned the white cliffs of Dover with a sign extolling Quaker Oats; he created half-ounce boxes, which he gave away by the thousands; and he even made a few unsubstantiated medical claims ("one pound of Quaker Oats makes as much bone and muscle as three pounds of beef"). He sold millions of boxes of oatmeal and through advertising became the true king of oatmeal.

The successful food entrepreneurs of the golden age of food processing were not distinguished by the fine quality of their products—though the first requirement of success certainly was to have a high-quality product—but by the creativeness of their advertisements. Swift, Armour, Hormel, Pillsbury, Dorrance, Washburn, Kellogg, Post, Heinz, and Crowell were all pioneers in creating advertising to make their products stand out from the crowd. The focus of marketing moved from claims of clean bowels and stronger muscles to associative advertising.

The most important association is and was trust. The consolidation of food manufacturers over the past twenty years, however, has made it difficult to determine the corporate pedigree of any food item. A current salsa television campaign derides the competitor as having been "made in New York City," when the brand being advertised is actually a subsidiary of the New Jersey–based Campbell Soup Company. Private labeling of food products—that is, placing someone else's label upon your product—has long been a common practice in the food industry. This was made quite clear to me on my first visit to a food factory, a tiny specialty canner in McMinnville, Oregon, where the jars of paté coming off the assembly line had six different labels. The cases of finished product behind me indicated that this small, specialty manufacturer was producing product under more than a dozen labels with a more than 200 percent range in the price of product from the same kettle.

TABLE 5.3 Select Food-Producing Conglomerates, 1996

Company	*Est. Sales ($M)*	*Select Brand Names*
Philip Morris (1866)[a]	65,125	Kraft, Post Cereals, Postum, Tombstone Pizza, Breakstone, Log Cabin, Entenmans, Lender's Bagels, Miller Brewing, Minute Rice, Jello
Cargill (1865)[a]	47,000	Commodities trading, flour milling, fisheries, Sunny Fresh eggs, fresh fruit juices, molasses
PepsiCo (1898)	24,472	Frito-Lay, Pepsi-Cola soft drinks, Pizza Hut, KFC, California Pizza
ConAgra (1919)	23,512	Hunt, Wesson, Healthy Choice, Beatrice, Armour, Swift, Eckrich, Oscar Mayer, La Choy, Chun King, Peter Pan, Banquet, Morton's, Patio, Orville Reddenbacher, and the company is the largest producer of generic chicken, and one of the largest of beef, turkey, pork, catfish, ocean fish
Coca-Cola (1892)	16,172	Coca-Cola soft drinks, Minute Maid
Sara Lee (1939)[a]	15,536	Sara Lee baked goods, Hillshire Farms, Kahn's National, Jimmy Dean, Superior Coffee, Bryan Foods
RJR Nabisco (1875)[a]	15,366	Nabisco crackers and cookies, Royal Pudding, Life Savers, Planter's Nuts, Grey Poupon
Anheuser-Busch (1875)[a]	13,734	Budweiser, Busch, and other beers, Eagle Snacks, Cape Cod Chips
Grand Metropolitan PLC	13,430	Pillsbury (1861), Burger King, Progresso, Van de Kamps, Pet, Häagen Dazs, Green Giant, Jeno's, Totino's
ADM (1923)[a]	11,374	Corn, grain, and soy products for industry (starch ethanol, syrup, flour, meal, etc.)

[a]Indicated company has significant sales in nonfood items.
SOURCE: Computed by author from multiple sources.

A complete list of the nation's largest food suppliers is impossible to compile, and the true level of activity of many is hidden, as they either are privately owned or are a small part of a larger entity. Concentration of suppliers has given them increasing clout with retailers to obtain space and prominence in the stores (see Table 5.3). Market control may be illegal in modern America, but shelf footage, shelf height, shelf location, control of end caps, in-store signage, and placement of products in store advertisements are all negotiable and all affect how the consumer perceives these giants. The giants are able to offer perks to stores that ensure better shelf locations. These perks include outright cash payments, better discounts on other products by the same manufacturer, in-store give-away programs, cooperative advertising programs, and the like. They may also see to it that some of their slower-moving items receive special attention through tying sales of these items to benefits associated with their faster-moving goods. The result is that the pattern has changed, as it has in so many other industries today.

The customer no longer walks into the store and says, "I would like five pounds of oatmeal and six bananas and some orange juice." The consumer chooses Quaker Oatmeal, Chiquita bananas, and Tropicana orange juice. These commodities have been successfully branded, the consumer pays a premium to obtain them, and the grocery provides extra perks to the manufacturer, who in turn provides additional marketing support.

The power of brand-name recognition was aptly established by C. W. Post as early as 1900. Postum was selling so well that a host of competitors entered the market with less expensive imitation ersatz coffees. Post ignored them initially. When their market share began eroding his profit margins, he counterattacked with a new, low-price grain coffee product called Monk's Brew. He sold the new ersatz coffee for a nickel a box rather than the quarter that was charged for Postum. Monk's Brew was so cheap that his inexpensive imitators were driven out of business in less than a year, but more important, the consuming public had had too much of poor-quality Postum substitutes. It refused to buy Monk's Brew as well. Taking back the Monk's Brew by the carload, Post had the boxes cut open and repackaged as Postum because it was indeed the same product. He reshipped the returned Monk's Brew as Postum and sold every box for twenty-five cents each. He felt vindicated when he turned his $46,000 loss in the first year of the campaign into a $385,000 profit the second. He should have been most pleased with the demonstration of the power of his advertising campaigns. People purchased Postum because of the image that he had been able to create, not because of what was actually in the box.

Distributors, Grocers, and the American Diet

The food-supply system has played a crucial role in the rapid broadening of the basic American diet over the past forty years. The fluidity of both the distribution and retail system has allowed the increasingly complex marketplace to create new demands and meet them comparatively quickly. Whereas there are some inherent problems in the ever increasing concentration of power of the distribution and retail systems into fewer and fewer hands, the very complexity of the market has meant that no single company or group of companies is likely to gain control of the production or distribution system, as occurred during the heyday of the robber barons. The Sugar Trust and the Tobacco Trust are only memories today and are not likely to return to haunt the American marketplace. There are too many sources of investment capital for any one entity to seize absolute control.

What consolidation has fostered is increased consumer power to obtain foods of higher and higher quality at more and more competitive prices. Distributors can guarantee growers that markets for exotic crops will exist; retailers can guarantee distributors that they have the market share to be able to distribute the volume of goods necessary to achieve economies of scale. The result is a greater variety of foods available at more reasonable prices than has ever been seen previously.

An analysis of trade-area market shares clearly shows that virtually all grocery retailers are regional in their activities. An examination of retailing through time further shows that if one of those retailers falters, the fickle market moves to another competitor. In the Atlanta market, for example, the traditional southern grocery chains were challenged for the first time by outsiders in the 1970s. They were eventually forced into secondary roles because their customer service (including their prices) had fallen below par. The Kroger Company attacked the market with bigger, brighter stores; more competitive prices; and aggressive marketing. Dominating the market, it too became lax, and Minneapolis-based Super Valu entered the market and became an almost instant force with its Cub Foods chain. Today both of these companies are stumbling as yet another competitor, the comparatively small but extremely well-managed Publix chain, enters the market. The fact that Publix is a much smaller company does not alter the fact that it has access to capital, which has allowed it to saturate the market with bigger, brighter stores. Noting that the current competitors were lacking in terms of service, Publix concentrated less on price and more on customer satisfaction. The days of customers shopping their entire lives at the Great Atlantic and Pacific Tea Company, drinking only Eight O'clock Coffee, and eating primarily Ann Page products are gone, at least for the moment. In most areas of the country the average consumer can purchase a wide variety of foods, and those not available in stores are available in catalogs and increasingly on the Internet.

The nationalization of the retail food-distribution system has also been largely responsible for the slow death of rural and small-town independent

Cherokee Grocery: Home of the Fighting Cherokee Indians

The narrow asphalt band of Texas Highway 16 winds across the eastern side of the Edwards Plateau and toward the center of the state. It was a bright, cool October afternoon with a little bite to the air. Drifting through San Saba County, we slowed for yet another town clinging to the edge of the road. Cherokee was little different than the others—three empty store buildings for every one that was occupied and only a handful of customers shopping in the entire community.

A low false-front grocery proclaimed itself as the home of the fighting Cherokee Indians, presumably a high school football team. Crouched at the crossroads, it seemed to be the liveliest place in town, though it had no customers. The Texaco station down the street had a parking lot littered with pickups, hoods up, but no one was around there either; the closed gas station directly across from the grocery was now home to a metal-puzzle company. Pulling around the corner onto the cross street, I parked the rental car and we wandered in for a midafternoon snack to try and kill the taste of the horrible Mexican lunch we had eaten an hour or so ago.

The store was almost an archetype of these dying businesses along so many roads in rural America. A false front, a rusting corrugated iron awning held up by even rustier posts, a crumbling sidewalk, three soft-drink machines representing each of the major cola manufacturers—Coca-Cola, Pepsi, and R. C.—an ice cooler, and a telephone. Deciding to see what was inside, we skipped the machines and went through the sagging screen door to be greeted almost immediately by the proprietress in the shadows behind the counter. Returning her warm reception and commenting on the weather, we assured her we would find our own way around the store. I made a quick circle through the canned-goods section, checking the stock. John, my indefatigable traveling companion, launched into a long discussion about beef jerky with the proprietress, who soon stopped worrying about two strange men casing her store.

A second round through the small, dark store demonstrated once again that there was nothing local, nothing unique, and nothing out of place in the entire store. The Frito-Lay man had recently been by, as well as the bread route man, who had stocked the shelves with plastic-wrapped breads and bakery treats. The canned goods were entirely mainstream and almost all in relatively small containers. Obviously the clientele either didn't use canned vegetables or they came here only when they ran out between visits to the supermarket in the next town.

(continues)

(continued)

Finding a coffin cooler in the darkest corner of the store, I reached down for a Diet Coke and discovered a strange bottle—Big Red.

"What's this, John?" I asked across the store.

"I had it down in San Antonio last year. It's a Texas brand."

"Any good?"

"It's an acquired taste," he commented with a small smile.

I scrounged around looking for a diet version but eventually picked up a bottle of regular soda and opened it while walking to the counter. Finding the fresh-fruit display on the way across the store hidden in a wooden cupboard, I looked it over carefully, amazed to discover that the locals must consume only fresh lemons, limes, and bananas. There was almost a full case of the limes. Someone in Cherokee must drink a lot of margaritas and cerveza. When I took a swig of the drink, a strange sensation moved down my throat that wasn't quite strawberry but was related to Dr. Pepper in some past life. It was not quite to my liking, but I continued drinking as I walked over to the counter and pulled a dollar from my pocket.

While the owner was making change, I looked over the signs tacked under the counter.

Its a boy! Paul Blayne Denton. 9 pounds, 8 ounces.
Proud brother Robby Denton.

San Saba County Christmas Parade.
Call now if you want to participate.

Bonfire and Stew Supper. Thursday, 7:00 p.m. Charley Perry's Place.
Bring dessert and cornbread.

The last one caught my eye. "What's that about?" I asked, noting that the event was that night.

"The football team is going to place San Saba tomorrow night and we are having a little get-together tonight to urge them on," replied the lady behind the counter.

"Never heard of a stew supper and bonfire. What kind of stew?"

"Deer meat."

"What else is in it?"

"Oh, you know, potatoes and onions and things. You're welcome to come on out and try it. Everyone will be there. You'll have a good time."

(continues)

(continued)

I looked at John, mentally calculating the time to Austin if we didn't get started until nine. He looked open but concerned. "No, but thanks, it does sound like fun."

"Well, you'll be missing a good time."

"I'm sure we will," John interrupted, "but we have to be in Austin tonight for a meeting. You have all this jerky; do you have any venison? My wife teaches kindergarten and likes to have an authentic Thanksgiving feast on the day before."

"John, they didn't eat jerky at Thanksgiving," I interrupted.

He ignored me, and she answered, "There is a place behind the gas station in Llano that makes the best deer-meat jerky around."

John thanked her and we both thanked her for the invitation and wandered toward the car.

"Why didn't you warn me about this Big Red?" I asked accusingly as went out onto the porch.

"I told you that it was an acquired taste," he replied with a smile.

groceries. The vast majority of these places have fallen on hard times as their numbers and stocks have shrunk. It has become popular to blame Wal-Mart for the disappearance of small-town business districts, but in reality these places were dying before Wal-Mart arrived. The continuing consolidation of the distribution system, coupled with the increasing size of the major grocery retailers and the competitiveness of the market, has meant that supermarket profit margins have been kept very low. There are no high-flying grocery equities on the stock market because profits come from extremely high volume with low markups. Small-town stores with low volumes have two problems. Many distributors do not want to deal with them, especially in produce, because it is inconvenient to sell in such small quantities. The prices of the small stores thus inevitably are higher than are those in the supermarket in the next town, which is drawing consumers from several communities. Declining sales volume in country and small-town stores has meant that they carry less and less of less and less. With declining total sales to amortize overhead, their stock stands out as relatively expensive. Typically there is less variety and the packages are smaller. Finally, these stores are often tiny, dark, and aging. As more and more local customers flee to the large, brightly lit supermarket in the next town, the local store slides to the status of an overlarge convenience store until the owner becomes too aged to continue in business. The fact that the store plays a role as a crucial community center at a time when the community is desperate for a sense of pride is lost in the swirl of day-to-day living.

C. W. Post, Huntington Hartford, and Clarence Saunders demonstrated that what we eat is very much in the hands of the processors and retailers. The public may not be coerced into eating anything, but it can be persuaded into trying almost anything. The result has been that tradition has been playing a decreasing role in what we eat and marketing has been playing an increasing role. Education is very much a part of this process and as such the cookbook in all of its various printed and electronic forms has played an increasingly important role in shaping our diet over the past century.

6

Cooking by the Book

There was a time when folks had cooks,
Who never did depend on books
To learn the art of cooking.
The help knew all the tunes by ear,
And no one dared to interfere;
They brooked no overlooking.

—***Charleston Receipts*, 1950**

The standardization of food preparation has been a problem since people first began mixing foods to enhance their flavor. Young girls traditionally learned their culinary skills from their mothers and grandmothers, and this pattern was little changed until recent times. Most homemakers had no use for a book of recipes, partially because they already had been trained to cook everything their families would willingly eat and partially because literacy was not high among women until the twentieth century. Whereas cookbooks have been around in America since the eighteenth century, most middle- and lower-class households did not have or use them. Neither of my grandmothers had cookbooks, though each had extensive libraries of other books; nor did my college-educated mother own one. There was a box of recipes in the cupboard under the counter. Those scraps of paper and printed recipes gleaned from friends and torn from *Sunset Magazine* and *Better Homes and Gardens* and taken from bags of flour and boxes of cereal provided guidance for many special meals. The bulk of all cooking in our home was based on recipes and methods passed to my mother from her mother and grandmother without benefit of written text. Lacking a daughter, my mother set about teaching me the ways of the kitchen as soon as I could reach the top of the counter. Perfect flaky piecrusts, German chocolate cake, spaghetti sauce, Sunday roasts, barbecue, and fried chicken were all in my repertoire long before I ever saw my first cookbook.

Early American Cookbooks

The need for cookbooks, of course, was always present in America, as large numbers of the nation's homemakers were immigrants far removed from their family sources of information about cooking. This demand increased even more rapidly with the beginnings of the Industrial Revolution; young girls were sent to the New England mills and factories to work alongside their mothers before any domestic training could take place. It is not surpris-

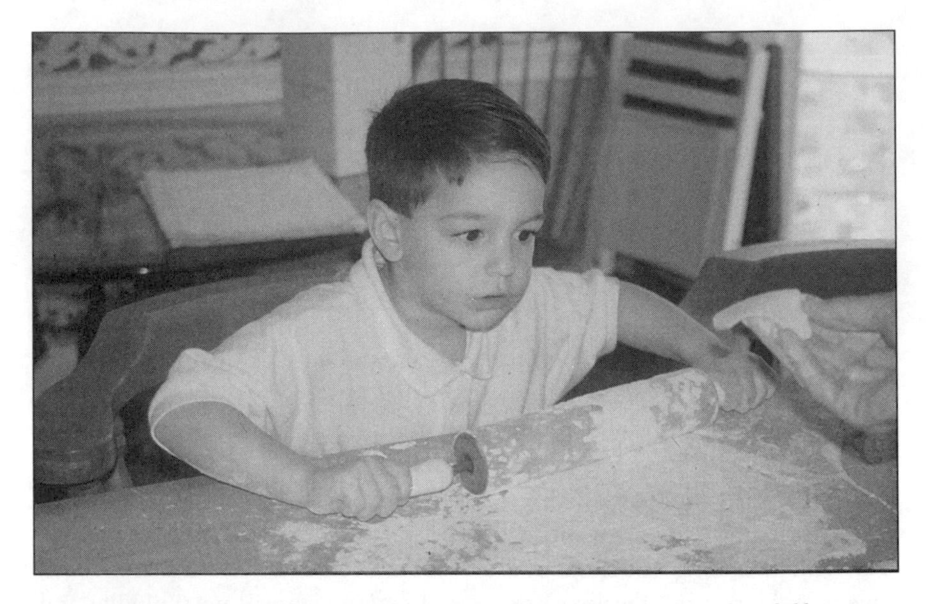

Learning how to cook was once an important part of a child's education; today, children are more likely to learn cooking at Grandma's house than at home. (Russell Weinwurm)

ing that the nation's first cookbook was published in New England, the home of the American Industrial Revolution. Factory women were probably not educated enough to use those books, but the wives of the factory managers and rising businessmen were. And soon they were wealthy enough to hire kitchen help, giving them even more time to try new recipes.

The first American cookbook was the forty-seven-page treatise *American Cookery, or the art of dressing viands. . .*, published in 1793 by Amelia Simmons. A variety of recipe books had been produced earlier, but they were little more than American editions of English volumes. Indeed, even Ms. Simmons's book owes a clear debt to earlier works like Eliza Smith's *A Compleate Housewife*, published in London in 1758. The first edition of *American Cookery* is especially interesting because it gives many insights into late-eighteenth-century cooking. In a later edition she disavowed an introductory section on the selection of foods that appeared in the first printing. She claimed that this section had been added by the printer without her permission.

American Cookery presents an image of food availability in those times that is in sharp conflict with the assumptions of most Americans. Recipes called for the use of a host of imported foods ranging from mangoes to Spanish (sweet) potatoes. Conversely, some stereotypes are contradicted. Apparently, the colonial fireside oven did not yield great warm loaves of yeast bread. Almost all of Simmons's bread recipes were for flat breads. Reliable commercial yeast was not available until the arrival of the Fleischman brothers more than fifty years later. Middlings could be used, but they varied significantly from batch to batch.

Other factors supporting the dominance of cornbreads and flat breads during this period were the lack of preservatives to keep the breads fresh after leaving the oven, the infrequency of bread ovens in most households, and the high cost of wheat flour. Bread without preservatives stays soft only a few hours after leaving the oven; yet baking took place on only one or two days a week. The lack of yeast-bread recipes, coupled with the large number of flat bread and cornbread products, suggests that few Americans had yeast breads on a daily basis. *American Cookery* was one of the first books to promote the use of pearl ash, an early form of baking powder, and the book has many recipes for biscuits and doughs for meat pies, the most likely breadstuffs consumed on a daily basis.

The book also reflects the tendency in most homes to cook only one full meal per day. The evening and morning meals either would be composed of leftovers from dinner or would be quite simple repasts. Most recipes call for enormous quantities of ingredients and lengthy cooking times. Missing are the recipes for the soups, potages, and stews one assumes were common during this period, and it is not clear whether they exist mostly in our imaginations or if they were so self-evident that recipes were not included. In either case, cooking on a wood fire in an open fireplace is much more complex than barbecuing chicken on the home grill. Cooking required constant attention, and few housewives of the time had all day to spend on cooking when the average family had many children, when washing started with the request to "go to the well and . . . ," and when the home garden, fowl cleaning, and many other tasks all awaited attention as well.

Simmons's recipes thus define a generally simple fare of plain meats, few vegetables, and a fair amount of breadstuffs. There are not recipes for any dish with the complexity of even a tuna-noodle casserole. It is obvious that the traditional twentieth-century dinner of one meat, one vegetable, one salad, bread, and a dessert was not a part of Simmons's world. Although a fair number of vegetable recipes are included, the lack of preservation technology meant that they were used only during the short New England summer. The remainder of the year the meat and breads were varied with turnips, parsnips, cabbage, and other storable cold-weather crops.

Mary Randolph's *Virginia House-Wife* is probably the second important cookbook published in this country. The book was first published in 1820; the following discussion is based on the 1825 (second) edition. *Virginia*

House-Wife was one of the first widely distributed cookbooks, especially in the South. It contains almost 450 recipes written in the traditional paragraph form. After a short soup section, including "Soup of any kind of old fowl," the book launches into meats. Slightly more than a third of the book is about meats and includes a lengthy discussion on meat curing. The meats are all cooked plainly, and much space is given over to the preparation of those parts rarely seen in grocery stores today—heads, feet, and organs. Only about 10 percent of the book is devoted to vegetables, almost a third of that to white potatoes, sweet potatoes, and (the nonvegetable) rice. Desserts occupy twice the space allotted to vegetables (although breads are included in this section). The book ends with preserves, pickling, and "cordials," a section that also has more recipes than the one on vegetables.

The interesting element of the recipes is the continuing dominance of British cooking. A handful of "Spanish" recipes—for example, an East Indian–style curry and a West Indian–style gumbo—are the only identified "foreign" recipes in the entire book. In contrast, quite a number of recipes that we would identify as African American today, including the mislabeled West Indian gumbo, are important in this book. The role of African Americans in the elaboration and regionalization of the southern diet is very apparent.

The monotony of the preindustrial diet may be the most striking picture painted by Mary Randolph's cookbook. More than a third of the volume is taken up with meat recipes, but the vast majority of those are for meat that is either boiled or roasted. Boiled beef, boiled fowl, boiled cod, boiled leg of lamb—the British influence on meat preparation was still strong in this society. The food was also quite bland; salt seems to have been the only universal flavoring. Occasionally thyme and parsley were called for in recipes, but most spices were rarely used. Cinnamon, mace, nutmeg, horse radish, and an assortment of others appear, but much less often than would be the case today. Whether this was a matter of expense or taste is unknown. Finally, the paucity of chicken recipes among a large number for other fowl reminds us that chicken was unimportant during the nineteenth century and that so many of our images of regional foods, such as southern-fried chicken, are not based on colonial or even nineteenth-century realities but rather are of comparatively recent vintage.

Changes with the Industrial Revolution

The arrival of millions of immigrants, the increasingly urbanized population, and the growing importance of processed and prepared foods called for the development of an entirely new approach to cookbooks during the late Industrial Revolution. The cookbooks of this period fall into three classes. Those sponsored by manufacturers to promote the use of their products were the most important of the new forms. Fairly uncommon at the time but a precursor for the future were those specialty books targeting individual regional or ethnic markets. Finally, there were the traditional general books, which became increasingly elaborate and important in shaping the nation's diet.

Companies such as Campbell Soup, Stokely's, Post Cereals, and Washburn-Crosby (later to be renamed General Mills) began producing proprietary recipes that blossomed into pamphlets and finally substantial cookbooks that demonstrated how old favorites could be made faster and easier with the new food products created by the publisher. Two interesting examples are International Silver Company's *What to Serve and How to Serve It: A Guide for Correct Table Usage* (1922) and *The Pillsbury Cook Book* (1914). The International Silver pamphlet specified the proper silverware and its placement for various social situations and included a sample menu and recipes for most dishes. The pamphlet was designed for the newly emerging and upward-aspiring middle class, which had little formal experience with place settings and multicourse meals but believed that such were important.

The Pillsbury Cook Book is interesting for its price, ten cents, and its completeness—more than 300 recipes with illustrations. The 115-page pamphlet comprehensively covered the same breadth of foods found in the more expensive cookbooks of the times. The editors chose to never specify company products in their recipes, resorting rather to a variety of direct advertisements of the company's products as well as more subtle ones extolling the high quality of its manufacturing process. Also included was a coupon for a free tour of the "A" Mill, the world's largest flour mill, and an order form for additional books.

The *General Foods Cookbook*, published in 1932, was far more blatant in its corporate self-aggrandizement. Subtitled "A Key to the Question of Three Meals a Day," the book never failed to specify the company's products by brand names in as many recipes as possible. A history of the company is also included as well as short histories and descriptions of its most important products. The reader is also very carefully tutored in why the company's products are always the best choice for the modern cook. Missing are recipes or mention of the company's very important frozen-fish product line. In summation, however, the book was comprehensive, was sold at a discount, and was written in a very matter-of-fact, easy-to-follow manner. The copy I consulted for this book was a 1932 Christmas gift to an Illinois housewife, who appreciated the book's assistance if the heavy wear is any indication.

A second class of corporate cookbooks was presented by publishers in the homemaking field. Butterick Publishing Company, a manufacturer primarily of dress patterns, began publishing pamphlets on food and cooking that culminated with the *Butterick Cookbook* (1911). *Ladies' Home Journal*, *Good Housekeeping*, and other "women's" magazines began publishing a few recipes in the mid-nineteenth century. The *Good Housekeeping* entry into the cookbook field was a logical extension of the increasing inclusion of recipes in the magazine. The book sold well, and the company quickly recognized that it was both an important source of income and a subtle advertisement for continued subscription to the magazine. Numerous editions appeared as quickly as it was believed that they could be absorbed by the market.

The thirteenth edition of the *Good Housekeeping's Book of Menus, Recipes, and Household Discoveries* (1922) is a good example of the standardization of recipes that had taken place by the 1920s. Shorter even than the *Virginia House-Wife*, this book still managed to cover the gamut of dishes because of its terse presentation style. Recipes were donated by readers to give the book the illusion of national coverage, though obviously the biases of the New York–based editorial team predominated. Whereas there were more spices and flavors, the complex dishes so common today were still comparatively rare. Further, it is interesting that there are far more recipes for lamb than chicken and comparatively little use of ground beef.

American middle-class families of this era typically had weekly meal regimes that varied little from week to week. Though the items might change from family to family, typically within each family the dishes for all but a night or two were standardized. In recognition of this phenomenon, most cookbooks like the *Good Housekeeping* volume included an annual round of menus with a week's menus for each month. It is doubtful that few actually followed this menu schematic, although there certainly were at least a few frustrated newlyweds who attempted to start life properly with these comparatively complex menus. Every breakfast featured two "entrées" as well as fruit; every main meal (midday on Sundays, evenings the remainder of the week) featured soup and dessert.

Books reflecting the cuisine of the new immigrants coming to America because of the Industrial Revolution also began appearing around the beginning of the twentieth century. *The Settlement Cook Book* (1903) was published by the Settlement Committee in Milwaukee, Wisconsin, as a part of its educational program for Jewish immigrant girls and women. The hearty Russian-German fare featured in this book felt comfortable to the largely German and eastern European population of the upper Midwest regardless of religion. Doughnuts, potato pancakes, rolled oats, bundt kuchen, pickled

Potato Pancakes

6 raw grated potatoes	*3 whole eggs*
A pinch of baking powder	*1 tsp. salt*
1 tbs. flour	*A little milk*

Beat eggs well and mix with the rest of the ingredients. Drop by spoonfuls on a hot buttered spider, in small cakes. Turn and brown on both sides.
(The Settlement Committee, 1903)

herring, and hasenpfeffer were all included to provide comfort food; baked corncake, oyster stew, and Boston brown bread were there to help the girls and women begin fitting into the new society. The book became popular throughout the immigrant Midwest despite the absence of pork recipes. Other regional and ethnic recipe books of the period included *El Paso Cookbook* (1898), *Clayton's Quaker Cook Book* (1883), and *The Kentucky Cook Book* "by a Colored Woman" (1912).

The market for cookbooks was vast, and these specialty books could no more than whet the appetite of the American public for books on sustaining a happy, healthy family. The increasing percentage of women working outside the home and the continuing high rate of internal and international migration meant that a high percentage of young families were without the benefit of a grandmother to help with the babies and teach those cooking skills that had previously been ignored. Guilt at not being home to cook the way one's grandmother and mother did surely played a role as well. The result was a spate of general cookbooks to help those who had not received home training in cooking. An interesting change in these books is the increasing inclusion of processed and prepared foods in the recipes; sometimes these were referred to by brand name, but more generic descriptions were given.

Both the overall number and the number of press runs of the cookbooks of the late nineteenth century increased dramatically, though they continued to be primarily purchased by the educational and economic elite. Catherine Beecher published the first cookbook that reached true national distribution in 1847, but it was Fannie Merritt Farmer who published the book that was to become so widely circulated that it continues to be reprinted today. After graduating from the Boston Cooking-School, she stayed on as assistant principal and later director. In 1896, a century after Ms. Simmons's first American cookbook, Farmer finished her landmark project. Her publisher was so skeptical of the potential of this 831-page volume that he required her to pay the initial cost of printing and to correct her own galleys. Despite his concerns, The *Boston Cooking-School Cook Book* was an immediate success. The first printing of 3,000 was sold by the end of that year, and the book was reprinted twice the next year, ultimately selling 71,000 copies before the appearance of the second edition in 1906. The second edition sold 455,000 copies in twelve years, the third more than 700,000 copies in five years. More than 2 million copies of the various editions were in print by 1941.

The *Boston Cooking-School Cook Book* was one of the first truly comprehensive cookbooks, though Fannie Farmer's predecessors at the school had published increasingly more complex books in previous years. The recipe list is distinguished by the continuing dominance of British recipes. A few recipes from western Europe were included, but virtually nothing from beyond that area. The book was essentially an attempt to ensure the status quo, not a program to lead the neophyte cook to new taste treats. Many recipes, such as fifteen for salt fish in the 1941 edition, were largely obsolete in later editions but were still included. Other changes in later editions included an increasing number of recipes featuring canned and processed foods. Despite this innovation, however, there was a continuing absence of foreign-food recipes in the early editions.

Fannie Farmer continued at the Boston Cooking-School until 1902 and then started her own school, which primarily was set up for housewives and society girls rather than professional cooks. She also contributed a page to *Women's Home Companion* with her sister, Cora Dexter Perkins, until Perkins's death in 1915. The Boston Cooking-School continued for only one more year after her departure, but its alumni continued to have great effect on American culinary life, most notably Maria Parloa, for a time part owner of *Ladies' Home Journal*. She also wrote the first cookbook for Washburn Crosby (later General Mills) and thus became the prototype for the fictional Betty Crocker. Betty Crocker was invented by Gold Medal Flour's Service Department in the 1920s to serve as a "persona" to answer the thousands of letters received with questions about the company's products and recipes. Betty Crocker was given a voice in 1926 when the company initiated a radio show and Marjorie Child Husted, one of the company's home economists, answered questions over the air. Betty Crocker became so successful that a competitor threatened to sue the company on the grounds of false advertising because there was no such person as Betty Crocker (Husted was thereafter selected to be the official individual representing the fictional homemaker). Betty Crocker has remained an important part of the company's advertising program, although both her look and her persona have changed greatly over the years. *Betty Crocker's Picture Cookbook*, first published in

1951, has become a fixture and general cookbook in its own right with little or no overt recognition of its corporate sponsor. It may be the nation's largest seller with more than 55 million copies in print.

Joy of Cooking, one of the nation's perennial best-sellers, also came out during this period and without any of the fanfare or corporate support that attended most of the other large sellers. Irma Rombauer, widowed at fifty-three, was encouraged by her grown children to write a cookbook to reenergize her life. The project was privately published in 1931 and sold about 2,000 copies. Attempting to broaden sales, Rombauer convinced Bobbs-Merrill to publish a new expanded edition. The 1936 version sold well, but it was not until the greatly expanded 1943 version appeared that the book became a kitchen bible.

The 1943 edition of *Joy of Cooking* included 3,500 recipes, half again as many as the revised version of the *Boston Cooking-School Cook Book* that appeared in 1941; more important, it was far more readable. Short comments are scattered throughout the recipes: "This is sweeter than the previous recipe" or "Rather luxurious treatment for this good bourgeois vegetable" give the reader a sense of comfort. Rombauer also used what has become the modern recipe style; ingredients are listed in the order they are used. The cook can easily make sure that all ingredients are available because they are printed in boldface type; yet their placement throughout the recipe allows the neophyte cook the opportunity to start the recipe knowing that items needed later can be processed as the dish evolves.

Opening the Farmer and Rombauer books side by side quickly reveals why Rombauer's book ultimately surpassed its earlier competitor. The recipes were more tuned to what mainstream Americans were likely to cook; the author's hints were helpful, not condescending; and there were many recipes geared to the faster pace of life that wartime America was encountering. Ultimately American cooks discovered that Rombauer's recipes worked, that her pie crusts were flaky, that her soufflés rose, and their families accepted the meals cooked from those recipes.

Foreign food continued to be almost totally absent in most cookbooks published between the wars. The intense anti-immigration sentiments of the nation were strong. The *Good Housekeeping* (1922) cookbook reflected those trends; virtually no foreign foods were overtly mentioned despite the presence of millions of citizens who had been born as aliens. The adjectives Italian, Arabian, and Bavarian appear, but they are just descriptive of the food, not attempts at presenting foreign food. Even such basic items in today's "foreign" lexicon as spaghetti, pizza, and pilaf obviously were perceived to be of little interest to middle America by the editors at that time.

Simultaneously, the renewed interest in America's past after 1920, which was reflected in the Rockefeller-sponsored development of colonial Williamsburg and the parallel rise of "early American" architecture and furniture, brought an increased interest in traditional regional foods. Boston baked beans, gumbo, and numerous other traditional regional dishes began peppering the pages of most general cookbooks. Ultimately, however, the most far-reaching change of the period was the inclusion of partially and fully processed foods and brand-name products into the recipes. By the end of the period, cookbooks were telling modern homemakers how to create whole meals in minutes by "doctoring" prepared foods into personalized cuisine.

The Modern Cookbook

It has been said that a cookbook is the closest there is to a sure thing in the publishing industry after the Bible. This may be hyperbole, but the volume and variety of cookbooks over the past fifty years has been staggering. Further, magazines targeting the home market typically divide their space equally among articles on home design, food preparation, and gardening. The result has been an avalanche of recipes and cooking tips that ultimately confuse as well as bedazzle the typical American home cook. A walk through a contemporary general bookstore reveals that more space is devoted to cookbooks than any other class of nonfiction.

Cookbooks also seem to have long lives. The *Fannie Farmer Cookbook* is now 100 years old and in its thirteenth edition. The *Joy of Cooking* remains a favorite as well, although both of these books have little in common with their first-edition ancestors. Another candidate for longevity is the *Better Homes and Gardens New Cook Book*. This was my first cookbook, given to me as a high school graduation present by a doting, childless neighbor. First issued in 1930 as *My Better Homes and Gardens Cook Book*, it has been a popular seller for more than sixty years, primarily because of the continuing popularity of its sponsoring magazine, which also gives its editors the resources to provide very slick presentations with lots of feedback from the market through letters to the magazine.

The 1996 edition includes about 1,500 recipes with hundreds of color photographs and color accents highlighting points the editors felt were important. Recent additions have been indications of preparation, cooking, and cooling times at the top of each recipe so that the harried homemaker can quickly find a recipe to meet his or her time restraints. Too, recipes are tagged as "low fat," "fast," and so on for the cook unable to make those determinations from looking at the ingredients. A list of the nutritional contributions of each standard serving of each dish has also been added. Finally, the ingredients are now not only listed separately from the descriptive text but appear again in the instructions in the order that they are added to the dish. If a complex dish with multiple parts, such as chicken and dumplings, calls for water more than once, the total amount appears in the ingredients list, and the exact amount needed for each use appears where appropriate.

The continuing popularity of this book suggests that a significant portion of the cooking public has little or no personal knowledge of cooking or even of the kitchen. This is not surprising considering the large number of dual-worker households, the rise in restaurant and take-out dining, and the tendency to send our youngsters to Little League rather than sitting them down at the kitchen counter and teaching them how to fry an egg. What is interesting is the book's selection of recipes. There is almost an international theme to the book with at least one foreign food appearing on each double page. In a casual perusal of the volume I found stir-fried tofu, Spanish rice casserole, meatless lasagna, polenta, and cornmeal mush on one set of pages and quesadillas, quick pizza bread, bruschetta, and cowboy caviar on another set. A careful comparison of these recipes to the same dishes described in specialty cookbooks, however, reveals a not-so-gentle Americanization of these "foreign" recipes—they are foreign, but not too foreign.

This trend, along with hundreds of ethnic cookbooks seemingly in every bookstore cookbook display, reflects the broadening of the American palate over the past twenty years, especially among middle-class Americans, who are the primary purchasers of these books. Truly exotic foods are few and far between; yet the neophyte cook is carefully led down the path of variety and increasingly nutritious meals.

Ultimately there are two types of cookbooks in today's marketplace: (1) those that attempt to provide basic information to the presumably untutored cook, and (2) those whose goal is to move the purchaser toward new kinds of foods and meal experiences. Books like Fannie Farmer's and the *Joy of Cooking* continue to be published in new editions for this market, as do those from other major publishing houses. These books typically contain few new ideas. Most purchasers utilize them to provide information on basic cooking—making a cake, baking biscuits, or roasting a turkey. More complex recipes are included but typically play a comparatively lesser role. All of us who cook a lot keep one or more of these so that when we forget the water-rice ratio or the roasting time for a turkey or are a bit foggy on a favored recipe, we can look up the information. Too, a fast comparison of *Joy of Cooking* and Craig Claiborne's *New York Times Cookbook* refreshes the memory before cooking the version that we learned while growing up in the kitchen.

The second class of books are targeted at intriguing the more advanced cook or gourmet. Many of these books are purchased by people who don't cook all that often but do enjoy reading about food. These books tend to be highly specialized and focus on a specific cooking style, cuisine, or dietary attitude. Books on barbecue, stir fry, casseroles, and the like are typical of those aimed at cooking styles. Books on ethnic foods, for example, Italian, Chinese, and American regional cookbooks, are even more popular. Less widely sold—except for those aimed at weight loss—but with extremely loyal audiences are cookbooks oriented toward specific philosophies such as vegetarianism.

Cookbooks and the Changing American Diet

It is one thing to suggest that American dietary tastes are rapidly changing and quite another to demonstrate it. Changes in the volume of consumption of constituent dietary elements are well documented by the Department of Agriculture. Little is known about the actual meals we are eating, but one approach to understanding the nation's changing cuisine is to examine the recipes printed in cookbooks and magazines (see Table 6.1).

The trends in American cookbooks from Simmons's 1796 book through 1996 well illustrate the changing patterns of food preferences. Simmons's book contains only 139 recipes and obviously has the least variety. The centennial edition of Fannie Farmer's cookbook (Cunningham, 1996), with more than 2,000 recipes, is the most comprehensive of the eight. The most obvious general food trends have been the increasing elaboration of the foods with more sauces and a greater variety of ingredients. The dominance of pork isn't obvious in the early cookbooks, though salt pork is present in dozens of recipes in each of the early books. It is clear in the early books that every scrap of every slaughtered animal made its way into the cooking pot; "variety meats" are not so frequently specified in modern cookbooks. There isn't a single recipe for an animal head in mainstream books after 1920, whereas there were several in most before 1900.

The rising importance of chicken in the American diet is especially obvious in reviewing these cookbooks, but less obvious is the parallel decline of other fowl. Not only are there more chicken recipes but authors assume that the cook is working with specific purchased chicken parts; earlier recipes in-

TABLE 6.1 Select Recipe Themes from Selected Cookbooks

	1796	*1825*	*1872*	*1903*	*1922*	*1942*	*1960*	*1996*
All fish	1.4	6.5	6.1	6.2	7.2	7.5	7.4	8.1
Beef & veal	4.3	5.7	8.2	3.4	3.9	3.4	7.8	4.0
Pork	0.7	4.5	2.9		2.2	2.0	2.3	2.5
Variety mts.	2.9	4.5	3.0	0.4	1.5	1.9	2.3	1.2
Chicken	0.7	1.3	1.8	1.2	1.3	2.8	3.3	3.9
Vegetables	5.8	8.7	7.2	3.2	5.5	6.5	7.5	10.6
Potatoes	1.4	1.8	1.3	3.0	1.1	2.1	2.2	1.8
Salads			0.4	2.4	6.4	3.9	3.4	5.0
Eggs		1.5	1.0	2.4	3.3	2.8	3.8	1.5
Pasta		0.2	0.3	0.4	0.4	1.0	4.0	1.7
Rice		0.3	0.8	0.2	0.9	0.7	1.6	1.2
Breads	10.8	6.5	10.8	9.2	8.6	5.0	4.6	5.7
Desserts	48.2	19.2	23.0	32.1	25.9	28.4	13.7	20.0
Preserves	12.2	5.5	4.0	1.6	2.4	1.8	1.3	1.5

NOTE: The books tabulated were Simmons (1796), Randolph (1825), Hill (1872), Kandor and Schoenfeld (1903), *Good Housekeeping* ... (1922), Farmer (1942), Truaxx (1960), Cunningham (1996).

cluded references to young birds, old birds, large ones, and small ones. There is almost a complete disappearance of other fowl—pigeon, squab, and duck—in modern cookbooks.

Recipes have become increasingly elaborate through time as more and more varied ingredients are routinely available to cooks in even the smallest communities. The changing recipes for gumbo reflect the kinds of changes that have taken place over the past 150 years. Mary Randolph (1825) probably published the earliest recipe for a dish she called "gumbo, a West Indian dish." It consisted of a pot of stewed "ochra" with salt and pepper and served with butter. Annabella Hill (1872) included several soups and boiled "ochra" recipes. Her gumbo recipe was more complex but still comparatively simple. A 1941 Fannie Farmer version of the recipe was even more complex. The seventh edition of the *Boston Cooking-School Cook Book* includes four gumbo recipes. The chicken gumbo recipe reflects a decline in the importance of okra and an increase in the importance of chicken. The dish is still quite simple, but the addition of tomatoes and parsley indicates that there is a general movement toward more flavor sophistication. The 1996 version of the same cookbook shows a significant alteration of the dish, as one would anticipate in the post–World War II era. Marion Cunningham's centennial version of the Fannie Farmer cookbook is a classic example of American postmodern cooking. Cunningham's updated recipe for gumbo is spicy and includes a variety of ingredients that were always available (and may even have been used in the real world of Cajun southern Louisiana) but that were never present in earlier recipes. Like so much of our culture today it is more authentic than the real thing ever was.

Magazines, in contrast to cookbooks, have consistently led the consumer in new culinary directions. Whereas cookbooks are purchased individually and the list of recipes is carefully examined with the assumption that they will be consulted over a period of many years, most magazines are purchased to help the reader stay current with changing trends and new concepts. Published recipes must be grounded in the past but also must be intriguing or have an interesting twist. Whereas some magazine recipes will be used over and over, most will be only perused or perhaps tried only once and forgotten when the following issue arrives.

This ephemeral quality of magazine recipes allows food editors to be more experimental in their offerings in the knowledge that their less realistic ones will be quickly forgotten. The key issue is that the recipes must appear interesting at first glance. As a result the very nature of the publishing process virtually forces the magazine food section to become a force for change in American cuisine.

An examination of a typical regional publication such as *Southern Living* illustrates the way in which magazines both reflect changing tastes and become agents for change. An examination of recipes from three years, 1970, 1981, and 1996, easily makes the point. *Southern Living* published its first annual-recipes issue in 1970 with 1,400 recipes. These were the most traditional and least innovative of all those examined. Expectedly, desserts were the single largest group, and they were not unlike those appearing in the general cookbooks of the period. There was a comparatively large percentage of recipes based on traditional southern themes with almost all of the region's signature foods represented—hoecake, hoppin' John, gumbo, jambalaya, greens, and the like. There were only a few "foreign" recipes, and those were heavily Americanized.

The 1981 *Southern Living Annual Recipes* book showed significant change in several directions. Recipes for the first time began showing the full impact of the prepared-food industry. Many recipes consisted of little more than mixing several prepared foods together in innovative combinations. The recipes typically had more spices and more new ingredients (especially vegetables), and more recipes were clear departures from old flavors and familiar ground. The magazine's audience was clearly being transformed from the largely rural original subscribers to the *Progressive Farmer* toward today's affluent, often suburban, educated elite.

Gumbo (1872)

2 pints young ochra pods
1 slice of lean bacon
a little lard
1 sliced onion
1 boned fried chicken
1 tbs. dried sassafras leaves
(Hill, 1872)

Chicken Gumbo (1941)

3 lb. chicken cut in pieces
1 1/2 c. tomato
1/2 finely chopped onion
3 c. boiling water
Salt and pepper
4 c. okra, cooked or canned
1/4 red pepper
Sprig of parsley
1 c. boiled rice
(Farmer, 1941)

Chicken Gumbo (1996)

1/4 c. olive oil
2 large onions
3 red peppers
2 lb. tomatoes (Italian plum)
2 lemons
1 tbs. fresh thyme
2 qt. water
salt and pepper
1 lb. Louisiana-style sausage
6 cloves of garlic
3 green chilis
1 lb. okra
2 bay leaves
1 tsp. fresh savory
2 tbs. filé powder
(Cunningham, 1996)

Traditional recipes continued in the 1981 edition; in fact, some recipes were essentially the same as those that had been published eleven years before, but they were clearly being replaced. The casserole seemed to have reached its heyday, much like the Corning Ware dishes they were to be baked in. These mixtures of meats and vegetables with a variety of flavorings were a common theme throughout the recipes of that year. Vegetables generally were more apparent, as were recipes with a foreign flare. Urban growth in Texas had made that state an important market, and for the first time southwestern and Mexican recipes began appearing in more prominent positions. Regional recipes tended to show more adventure generally. Typically, the opening photographs of the 1981 book of recipes are of a pot of hoppin' John and another of collards, but the back cover included Grilled Asian Chicken Salad and Herbed Tomato Tart (a pizza-like dish).

If the recipes published in 1970 represented the traditional South, those published today are a clear reflection of the new South. Not a single recipe for spoon bread or hoe cake appeared in 1996, only seven recipes for grits, and two for hoppin' John, one of those a salad. Traditional recipes have been replaced by more appetizers, more fish, more salads, more vegetables, and more foreign foods, especially those with an Asian flair. Fat, the basis of the region's traditional cooking according to its critics, had become a dirty word. Traditional southern ingredients—shrimp, oysters, greens, and cornmeal—are still a familiar part of many recipes, but they are being used in new ways. Examples of the juxtaposition of the past and present include Crawfish and Tasso Fettucine and Crab Cakes (sautéed, not fried) with Greens and Dijon. The term "nouveau southern" is never used in the magazine's recipes, but there is no question that its editors are helping readers, many new to the South, to find ways to continue traditional regional favorites without the grease and long cooking times that characterized the food of the past.

The same kinds of shifts have taken place among *Southern Living*'s many competitors, both regional publications such as *Sunset* and *Yankee* and national ones such as *Good Housekeeping* and *Better Homes and Gardens*. If the role of magazines is to help the reader be current and have at least a glimpse of the future, then the recipes must do so as well. These magazines, more than ever before, have become arbiters of national and regional tastes. *Sunset* magazine has long been a leader in residential architectural design on the West Coast, and *Southern Living* (now owned by the same company) has followed its lead in the South. These magazines select house plans and gardens that they believe exemplify their regions and showcase them in articles to spread the *Sunset* and *Southern Living* gospel. Developers are selected to construct particular plans, which are then open to the public for a fee and advertised in the magazines. The sale of house plans has become a major business of both magazines as they attempt to define their regions in contemporary terms. Similarly, both magazines use their position as taste arbiters to influence the choice of foods served in the trade areas.

A second important source of new cooking ideas over the past thirty years has been television cooking shows and demonstrations. Cooking demonstrations have long been a part of television, especially in the early days, when local stations were left with hours upon hours of broadcast time to fill during the day. In 1961, Julia Child appeared on a WGBH (Boston) book show to talk about her new book *Mastering French Cooking*. The producers liked her. WGBH is one of the most innovative public television stations in the country, and *The French Chef* went on the air in February 1963.

She became an American cooking icon for her easy way of explaining cooking and for her many faux pas, which were broadcast in those days of live television. Most Americans cook little more French food today than they did before they began watching her many television shows, but her impact has been tremendous. One of Child's most important contributions to the evolving American cuisine was a willingness to stand before a camera undaunted by her spills, splashes, and failures. She demonstrated that fixing problems is a part of cooking. Ultimately, her style encouraged home cooks that they *could* cook French or Italian or whatever and that the result would taste good. Like Leonard Bernstein with his children's concerts of the same period, Child demonstrated the basic structure of an apparently complex process in terms that any American could understand.

Julia Child was joined on public television by many others demonstrating a host of exotic and regional cuisines including Louisiana Cajun, Italian, and Chinese. These cooking shows demonstrated more specifically than did cookbooks how to create new dishes, showed what they looked like, and always ended with the host and some visiting victim sampling the day's output while smiling and saying, "Delicious!" Traditional printed cookbooks did evolve out of almost all of these shows; Justin Wilson, Jeff Smith (the Frugal Gourmet), and dozens of others became household names. Even though daily cooking has been increasingly taken over by restaurants and manufacturers of prepared foods, complex cooking for special occasions is becoming popular and is more sophisticated than ever because of these information sources; they have been largely responsible for the creation of today's polymorphic cuisine.

7

Imported Tastes: Immigration and the American Diet

First Generation: Bagel and lox with a glass of tea.
Second Generation: Bagel and lox with a cup of coffee.
Third Generation: Bagel and Nova Scotia salmon with a cup of espresso.
Fourth Generation: Two croissants, an omelet aux fine herbes, and a glass of skim milk.

—Hochstein and Hoffman, 1981, in describing the acculturation of the Jewish breakfast in New York

America has long described itself as a melting pot. The culinary pot too has absorbed the flavors of the millions of immigrants of hundreds of ethnic backgrounds who have settled here. In 1994, there were 23 million foreign-born residents living in the United States; more than 56 million emigrants have come to these shores (see Figure 7.1). All Americans are immigrants or descendants of immigrants. The ebbs and flows of these people with their divergent cultural histories have shaped the nation and its distinctive cuisine. A short immigration history of the nation thus is important to understand the changing character of the national diet through time, as well as its continuing regionality.

Connections between people and place are quickly established; migration tends to involve concentrations of people who have the same origins settling in only a handful of destinations. Some emigrants may have purchased tickets to America without thought to where they might end up, but few had the temperament to randomly pick a destination from a steamship or railroad schedule. Decisions were for the most part guided by knowledge of the destination, although certainly my grandfather, initially on his way to the Oregon Dales, had never heard of Chico, California; someone he met on the ride west convinced him to get off the train in Oroville and buy land in nearby Chico. Most of us move to connect with family or friends. In colonial times many towns and villages were inhabited by people of similar backgrounds; in the nineteenth century the trend became even more pronounced, as people with common goals organized to migrate as a group. There are numerous examples of communities in the Midwest initially populated almost entirely by people from a single village or district in Scandinavia, as there are mining towns in Pennsylvania with similar links to villages in the Ukraine and Poland and even earlier Puritan villages with most citizens coming from a single village or county or two in England.

Little has changed in the nature of migration streams, though origins and destinations have altered. California is now home to almost one-third of all

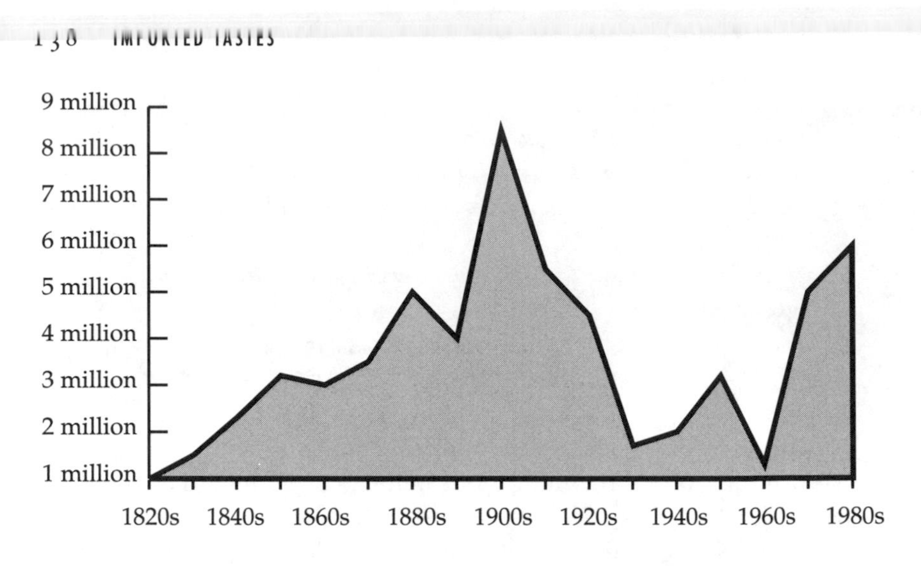

FIGURE 7.1 U.S. Immigration by Decade: 1820–1989

foreign-born residents, followed by New York with about another 13 percent and Florida with just under 10 percent. Newly arriving immigrants feel more comfortable living in communities where people speak their language and where the foods and music of home are readily available.

On a microscale, immigrants typically select areas within cities where they will find people of similar backgrounds. Certainly part of this selection process is affected by lingering housing discrimination, part stems from the economics of arriving in a new country with inadequate resources, and part relates to accessibility to the kinds of jobs available to immigrants with limited English. There is also an emotional need to live in a community of people who seem familiar, who have something in common with the people left back home.

There are new immigrant districts in all of the nation's larger cities today. Unlike earlier ethnic ghettos, these districts tend to be sprawling suburban or semisuburban areas of small, inexpensive, single-family homes, strip malls, aging apartments, and bustling multilane streets. Ethnic mixing still takes place in all but the largest of these zones, which may be inhabited by as many as thirty different nationalities in a few square miles. The result is a mélange of cultures and lifestyles and constant adjustments.

Atlanta's largest immigrant district, for example, sprawls along a strip of apartments and inexpensive, single-family houses stretching almost ten miles along Buford Highway. Immigrants from so many nations live along this street that a dozen different native languages may be spoken in a single elementary school classroom. Although people of more than fifty national origins live along this street, the single most distinctive element is two subneighborhoods of Mexican immigrants living at opposite ends of the strip. Those on the south end proudly advertise their origins from Jalisco with signs and flags on their cars. Those on the north end are just as proud to have emigrated from Michoacan. Each has created an independent sense of community with its own dance hall, restaurants, and grocery stores; yet the traditional single-culture ghetto has not come into being.

These are not ghettos; open-housing legislation has led to more than a dozen smaller ethnic areas scattered throughout the city, and tens of thousands of immigrants live totally integrated into mainstream Atlanta life. This larger district came into being because of the thousands of inexpensive apartment units and houses lying along its route and continues because it is well connected to the jobs that these immigrants typically hold. Most residents will stay only a few years before they leave for better housing elsewhere, though few will remain totally unconnected to this place as long as they live in this city.

The major grocery chains still maintain stores here, but their market share is less than elsewhere in the city because they do not stock the specialties in demand in this constantly changing cultural mix. Small family-owned bodegas, oriental groceries, and specialty bakeries have all opened to offer their countrymen authentic foods from home. A few years ago a warehouse a few blocks away was converted into a sprawling international grocery serving the entire community.

Dining out in a restaurant along this great immigrant commercial strip is an experience to remember. One does not go to a Mexican or Chinese restaurant; rather one chooses among Moroccan, Chihuahuan, Szechwan, or a dozen other specific cuisines. The menus of many of these places are slowly Americanizing—rarely does one see menus only in Chinese or Korean characters anymore—and almost all offer a few mainstream favorites to attract a larger clientele. Overall, the change process is little different from what took place among the Slovak coal miners in Pittsburgh or the Polish meat cutters in Chicago a century ago.

A Short Immigration History

The first American immigrants were the descendants of foragers who drifted out on the Bering Sea land bridge during the late Pleistocene and then drifted southward through the generations into what is now the United States. Their

Instant Masa Mix and Sangqua Okra

The International Farmers Market is a low-lying yellowish building across the road from the Red Baron Antique Mall and Auction Gallery and underneath the final approach pattern for Peachtree-Dekalb Airport, one of the busiest general aviation airports in the nation. A once-abandoned warehouse home of a discount mattress outlet, this unpretentious building now serves the growing in-town international community as a food emporium offering specialties available nowhere else in the city. With 80,000 square feet for foods on display and 20,000 more for storage, the I.F.M., as it likes to call itself, doesn't carry wheat thins or Campbell's soup, but if you want cactus leaves, 100-pound bags of pinto beans, Jasmine rice, or Virgil's Micro-Brewed root beer from England, this is the place to look.

It was obvious that it was Saturday when we pulled into the parking lot. Cars were parked hither and yon in every available space and overflowed into the rapid-transit lot next door. Five or six green, white, and red taxis, many boldly proclaiming "We Speak English" on their sides, waited at the door to take shoppers back to their apartments along Buford Highway. It was a mob scene but also the only place to shop for the dinner we had planned for this cold February weekend to break the monotony of winter.

As we worked our way through the people in front of the doors, the crowds closed in on us like a good-natured wave. Carts jostling, children running between one's legs, people shouting and pointing—it could have been a market anywhere in the world except that no two people seemed to be speaking the same language. Working our way through the bread displays—maybe a hundred different kinds from Near Eastern flat breads to European round loaves to tortillas—we arrived at the fresh-fish counter. The character of the customer base became clear. Certainly there were two sizes of squid and octopus, but catfish,

(continues)

contributions to the American diet were slight, but some of their descendants settled in Mexico, where they created an agricultural economy based on indigenous crops. These people later expanded back into the United States as agriculturalists and did have a major impact. Their new agrarian economy spread across most of the eastern United States and portions of the Southwest before the first European settlers arrived on the East Coast.

The classic Amerindian diet at that time focused on the consumption of corn, beans, and squashes supplemented by varying quantities of chilis, peppers, and other domesticated and nondomesticated vegetative foods. The

(continued)

mullet, carp, and bream were there too. This was not entirely an international market, and that ten-pound pail of chitterlings in the meat case was not there by accident.

We finally made it clear to the counterman that we were looking for fresh tilapia, and he directed us around the U-shaped fish area to some tanks on the other side. Passing the mullet, catfish, and mudfish, we discovered that "fresh" in this market meant still swimming. The counterman grabbed a net and pulled out a couple of likely candidates amid a flurry of splashing that got both us and him wet—these were not half-dead tank fish. We selected a couple, and while they were being filleted we drifted over to the vegetable department.

Forgetting our mission when seeing the variety of vegetables I did not recognize, I began reading my way down the aisle. There were thirteen vegetables in bins between where I was and the next aisle. I knew daikon but wasn't aware that it came in Korean and Chinese versions. There were six kinds of squash and seven kinds of melons; I couldn't even pronounce the names of some. I discovered that "European" vegetables were two aisles over and headed that way, though first I grabbed a few peppers and a small winter melon.Working our way back through the buggies, we headed for the rice section. Hundred-pound bags of basmati, long-grain, short-grain, pink, Thai, and Indonesian rice sat on the floor, snuggled up to the fifty-pound bags of Lundberg Family organic rice from California. Incongruously among the large bags was a counter of six-ounce envelopes of Grey Owl Wild Rice. Grabbing a moderately small bag—ten pounds—of short-grain, organic, brown rice I headed back across the store to pick up the fish, which was now ready. Spotting open space at the front of the store, we dashed toward the cash registers and outside, hoping we had everything. If we didn't we would have to fake it, because there was no way we were coming back on Saturday morning next week.

subtropical crop elements declined in importance in the northward migration until they disappeared; they were replaced with some local domesticates and increased quantities of game.

The first British colonists on the East Coast set about replicating their traditional diet, which remains the foundation of the national mainstream. In the short term the British immigrants were also forced to accept most of those crops that had been introduced earlier by the Amerindians, but the actual diet changed little. Corn was well suited to the environment of most of the eastern colonies; the varieties of wheats and barleys brought from England were not. Cornmeal became a staple of frontier life throughout the eastern United States, but even that disappeared from regular use as more traditional wheat flour became increasingly available through time.

West and central Africans were the third major ethnic group to enter the American culture during the early colonial phase. African Americans were not voluntary migrants like most of the Europeans; the first shipload of African slaves was brought to Jamestown, Virginia, in 1619, and thousands followed over the next 189 years until the importation of slaves was banned in 1808. More were illegally imported after that date. Virtually all African Americans today can boast of American heritages lengthier than their European neighbors. African-heritage Americans accounted for about 19 percent of the nation's population in 1790; that number has slipped to less than 10 percent today. Virtually no Africans voluntarily emigrated to America prior to World War II; only in the past several decades has a small trickle of African migrants arrived as the nation has striven toward legal, if not actual, equality of races.

African influences on southern cuisine were monumental and generally were not acknowledged until recent times. Whereas slaves were provided with basically European foods by their owners, most were able to supplement those rations with foods grown around their living quarters. The most important of these crops included sweet potatoes, peas, peanuts, watermelon, and in some areas rice.

Significant regional variations in the foods cultivated were apparent almost from the beginning. Rice growing initially was concentrated in the coastal zone stretching from modern Georgetown, South Carolina, southward to Darien, Georgia, but ultimately stretched along the Atlantic coast from North Carolina into northern Florida. Sweet potatoes were grown throughout the Atlantic and Gulf Coast South but were more important as a food supplement on the Gulf Coast. Peas—various kinds of chickpeas, black-eyed peas, and similar beanlike legumes, possibly originating in the Middle East and introduced into West Africa by Arabic traders—were widely consumed throughout the entire Lowland South. Peanuts, okra, watermelon, and a host of lesser crops were also widely distributed. Some African-origin sorghums may also have been grown, though sorghums were not "introduced" into the region until the late nineteenth century. A host of African-origin dishes like hoppin' John and gumbo were created from these ingredients, although only recently has research in West Africa begun to clearly establish their geographic ties.

Western European groups dominated the remainder of the early immigrant groups settling in what is now the United States. Small colonies of

Yoo-ke-Omo (A Ghanaian Hoppin' John)

1 c. cooked black-eyed peas
Omo (see below)
1 tsp. ginger

Use stock from cooked beans and some water. Stir in ginger powder and beans when butter is added. Excellent with meat or fish dishes.

Omo

4 cups water
3 c. uncooked long-grain rice
1 level tsp. salt
1 tsp. butter

Bring salted water to boil.

Wash rice three times in cold water and add to boiling water. Add butter. Bring to boil again immediately. Stir once, then cover with a tight-fitting lid. Reduce heat to very low and allow to cook slowly, undisturbed, for 20 minutes. Turn heat off and remove from burner.

Without lifting lid, allow rice to steam dry for another 10 minutes. Serve while rice is still hot.

(adapted from Ayensu, 1972)

Dutch and Swedes settled in the Hudson and Delaware River valleys during the seventeenth century. Swedish settlers along the lower Delaware never numbered more than a few hundred and had little impact on the culture generally, although the question of a Swedish origin for the American log cabin has again been raised in recent years. Early Dutch settlements were larger, numbering about 80,000 inhabitants along the Hudson River and in northern New Jersey at the time of the first American census in 1790. It is difficult to ascertain the influence of the Dutch on American foodways because of the later, larger migrations of similar peoples into much the same regions. The *wafle* is the most commonly cited Dutch food item widely consumed in America then or now. It is also known that the Pilgrims acquired knowledge and appreciation of this dish when they lived in the Netherlands prior to coming to America. The dish could just as easily have been introduced by the Pilgrims, the Dutch, the Scandinavians, or the Germans and probably was introduced in some areas by each of those peoples. There has been a rebirth of Dutch food along the Hudson Valley in recent decades, but that has more to do with the general revival of regional menus and place associations than with a continuation of family traditions.

William Penn's liberal Quaker attitude that everyone was welcome in his colony attracted a wide variety of immigrants, most notably large numbers from central Europe, nominally called Germans or Pennsylvania Dutch. These peoples primarily entered the United States through Philadelphia, Wilmington (Delaware), and Baltimore, settling west of those cities. They developed one of the most distinctive colonial ethnic homelands in America, though only remnants of it still exist today. Their diet too took on a character of its own and is discussed in the previous chapter. Later Germanic immigrations into the United States, however, tended to jump these rural settlements and had little impact on the Pennsylvania Dutch way of life, though large numbers of these later immigrants did settle in Philadelphia and Baltimore.

Immigration to the United States during the late colonial period and the early nineteenth century was a relatively small flow. Accurate statistics of annual in-migration are not available for the colonial period, but only about 40,000 migrants a year entered the country during the 1820s. Many of these were from Germany and Ireland. New York was the largest port in the United States, and the shipment of heavy natural resources to European ports meant that return passenger traffic on the empty ships was relatively inexpensive. Two-thirds of the German immigrants came first to New York in the 1840s; most of the remainder took advantage of the low fares offered by empty tobacco ships returning to Baltimore or cotton carriers on their way to New Orleans.

Many of the New York arrivals from ports in Germany remained in that city or its environs. Others either traveled by rail into upstate New York, especially Buffalo, where a large German ethnic population still resides, or continued to the large industrial cities in the Midwest. Those entering other ports tended to head toward the midwestern frontier, where strongly Germanic-origin populations are still present in a wide band from Cincinnati northwestward into southern Minnesota. They typically did not live in nearly as cohesive communities as the earlier Pennsylvania Dutch, but many areas within that band continue to be identifiable as to their ethnic roots. St. Louis, Cincinnati, and Milwaukee all have large, influential German communities; most other industrial cities had somewhat smaller numbers. A large group of Germans also settled in the Texas hill country after 1831.

Today, it is difficult to separate the influence on American cuisine of this large mass of later Germanic settlers—primarily concentrated in the nation's agrarian heartland—from that of other northern European settlers. The higher consumption of pork in the rural Midwest is one of the most obvious reflections of the Germanic stamp on early settlement, as is the generally

Shof-Fa Noodles

Frozen bread dough
Raw potatoes (diced)
Cold water
Onions
Salt and pepper

Let the frozen bread dough thaw until you can cut off enough for one dumpling, then shape and tuck in ends. Make about 10 more, enough for a large deep frying pan. Let dumplings rise until double in size. Put Crisco and onions in frying pan. Next add the dumplings and potatoes. Salt and pepper to taste and pour cold water over (almost covering) about 1 c. Keep lid on while cooking and do not open. Let fry 20–25 minutes or until golden brown and crispy.
Mrs. Al Meidinger
(Max Legion Auxiliary, 1974)

heavier midwestern rural diet. The later German settlers also brought a potato-consuming tradition, as did the Irish and others.

A variety of widely specific popular food items, virtual icons in the American diet, certainly are ascribable to German influence. The all-American hot dog may have been invented here, but its constituent elements, most of its early progenitors, and even the technology to make the all-important wiener are all of Germanic origin. Other contributions to the contemporary American diet include the use of potatoes to create a salad, the egg noodle, and the holed doughnut (Mariani, 1983). The popularity of cabbage, specifically coleslaw, is more difficult to ascribe to a single ethnic group. The term coleslaw probably originates from the Dutch *koolsla* (cabbage salad), and the first use of the term dates from 1792. As such it may be of Dutch origin, but the early widespread consumption of coleslaw, coupled with the variety of recipes, suggests a Germanic hand in its rise in popularity in America.

Lager beer may have been the most visible of the Germanic culinary contributions to the American way of eating. Many nationalities brought beer traditions with them to America. The first brewery in the United States was constructed by Dutch settlers in New Amsterdam in 1612. The Pilgrims were also beer drinkers; indeed, the Mayflower's log suggests that it was the depletion of the ship's beer supply that played a crucial role in the decision to call off the search for Virginia and settle in Massachusetts. The first brewery opened in Virginia in 1629 and the first licensed brewhouse in Massachusetts in 1637. English beers typically were heavy-flavored, dark beverages; German beers, often derisively called "women's beers" by the English, were lighter.

The lighter German lagers ultimately displaced the heavier British products as public preferences changed. The beginning of the end dates from the 1840s, when a German brewer in Philadelphia began selling a new, lighter, bottom-fermented lager beer. Lager and the lighter pilsner beers (brewed by a Czech process that was popularized in this country by Germans) became the American standard by the mid-nineteenth century. Philadelphia was the nation's first great beer-producing city, though New York had its share of famous mass-produced brands as well. The Schaefer family of New York City purchased a brewery in 1842 and made it one of the most popular East Coast beers. Milwaukee, a center of German immigration, rapidly expanded as a beer center after the Best family arrived from New York and built a brewery. Frederick Pabst purchased the brewery a few years later and sold more than a half-million barrels of beer in 1889.

There were 4,000 breweries in America in 1876, certainly not all owned or operated by Germans but still heavily influenced by their beer tradition. British stouts, porters, and ales have always been available in America, but for more than a century, the standard mass-produced beers have been the Germanic lagers. Whereas the largest centers of per capita consumption throughout most of the nation's history were in Germanic settlement areas, consumption today has little to do with ethnicity.

The impact of the Irish on American cuisine is difficult to measure. Commercial potato production was successfully begun by Scotch-Irish settlers, but so many early immigrant groups consumed potatoes that it would be difficult to credit any single group for that food. Ethnic historians like to suggest that the Scotch-Irish were responsible for the rise of whiskey in the United States. Certainly some Scotch-Irish did distill grain beverages in the eighteenth century, but so did numerous other groups. The demand for distilled beverages was so widespread and the technology so simple that, again, it is impossible to ascribe the innovation to any single group.

The flood of Catholic Irish after the potato famines in the early and mid-nineteenth century created renewed demand for potato production but otherwise had little visible impact on the national diet. Most of these immigrants were poor, settled in cities, and had little power to alter anything in their new homeland. Mulligan or Irish stew has an honored place in history, as does corned beef and cabbage, but the dishes have typically had only a small place on the American dining-room table.

The Early Industrial Revolution: Western and Northern European Emigration

The post–Civil War era brought renewed immigration to the United States as the Industrial Revolution brought widespread economic readjustments to Europe. Many commercial European grain farmers on the agricultural periphery found it increasingly difficult to continue farming at a profit. Faced with tough choices—change their crops to increase profitability, migrate to the city, or slowly slide into economic decline—many chose to leave Europe.

Economic unrest in northern Europe was accompanied by religious unrest. The first Swedish immigrants to the United States in the nineteenth century were religious refugees who settled in Illinois in the 1840s. The largest group established a utopian communal society in Henry County that lasted for less than twenty years. Others came for the cheap agricultural lands. Most Swedish farmers from this period arrived with sufficient money from the sale of their farms back home to buy farms in settled areas. Later Scandinavian settlement was less and less well financed, forcing settlement further north and westward toward the agricultural frontier and less expensive lands. Minnesota and Wisconsin were the initial centers of Scandinavian settlement; large numbers of later settlers were found along the rail lines in western Minnesota, the Dakotas, Iowa, and Nebraska.

The concentrated nature of Scandinavian settlements in the peripheral northern Midwest and later the Pacific Northwest tended to minimize their impact on the national cuisine. Whether by design or accident, many communities were heavily of one ethnic group, often from a single section of Norway or Sweden or even Iceland. The result was the development and continuation of strong ethnic ways of life outside the Twin Cities and other major centers feeding immigrants to the American grasslands.

Carrie Young captures the essence of these Scandinavian communities in her charming reminiscence cookbook *Prairie Cooks:*

Lutefisk og lefse
Gammel ost og prim
(Lutefisk and lefse
Old Cheese and whey)

For her father, the ingredients mentioned in this couplet, in addition to a little rommegrot, were essential. These basics were accompanied by American-style food: a small steak and fried potatoes for breakfast, meat and potatoes for the noon meal, and a bowl of tomato soup and homemade bread for supper (Young and Young, 1993).

As in many ethnic communities, in the Scandinavian ones it was only a generation or two before the vast majority of the food consumed became mainstream American in character. There were subtle adjustments: a bit more fish, often local whitefish or salt cod, many more pancakes of all kinds (especially with potatoes), and far more desserts—cakes, pies, and cookies for the most part. Traditional foods, often labor intensive, begin disappearing from the weekly menu until finally they were consumed only on holidays. Christmas, Easter, and Lutheran church holidays were all celebrations of the past; meals on those days included lefse, lutefisk, and the other specialties.

The individual dishes, whether Norwegian or Swedish or Finnish or Icelandic or Danish, seem much the same for those who are not a part of the community. A dried-cod entrée, lutefisk for the Norwegians, is usually the most prized of the main dishes. It is first soaked in water and then a lye water and finally simmered for a few minutes and served with clarified butter. Flat breads, most often pancakes in this context, are a second item with strong regional identifications. Much thinner than the American versions with baking soda, these flour or potato and flour flat breads are highly prized and surprisingly difficult and time consuming to make correctly considering their simple ingredients. A typical lefse recipe has six to eight pounds of mashed potatoes to one cup of flour. A combination of milk, cream, and butter is the binder, though this bread is very fragile until baked on a griddle. Finally, there is an assortment of cheeses, varying with the region of origin in Europe, and, of course, desserts. Large quantities of cream and butter and comparatively small quantities of flour tend to make the cakes and cookies light and fragile.

The Late Industrial Revolution: Eastern and Southern European Emigration

More than 13 million Americans today claim to have single national ancestry from within central and eastern Europe. A million or so of these are the descendants of post–World War II immigrants; most are the children and grandchildren of turn-of-the-century immigrants. The Poles were the first of these groups to arrive in large numbers but were soon followed by a host of Czechs, Slovaks, Ukrainians, Lithuanians, Russians, and others. Many

Grandma Jensen's Aebleskiver (Round Danish Apple Pancakes)

2 1/2 c. flour	*1/2 t. salt*
1 1/4 t. baking soda	*3/4 t. baking powder*
3 eggs	*2 c. buttermilk*
3 T. butter	*diced apple to taste*

Sift dry ingredients. Add beaten yolks and buttermilk, then melted butter; fold in beaten eggwhites. Drop into light greased aebleskiver pan, filling the sections half full. Sprinkle with cinnamon and drop in a few small pieces of diced apple (peeled). Cook a few minutes on medium heat—then gently roll them over to cook the other half. It takes 3 to 5 minutes per side, depending on type of stove. Serves 4–6.

Ann Schippman
(Edison Park Lutheran Church, 1978)

ethnic historians distinguish Jewish immigrants as a separate subset of these groups even though that classification is based on religious grounds rather than ethnicity.

Most of these groups left Europe with few skills that could be applied to an industrial society, and many of them had to accept the worst jobs of the time—in coal mines, steel mills, and construction—which required brute strength and stamina rather than skill. Most could not take up farming because of a lack of capital to purchase land, though there were some notable exceptions among several small groups of Poles. The vast majority immigrated to the industrial centers of the nation, and the current distribution of these national groups still reflects that bias. The largest centers are in New York City and Chicago; lesser numbers reside in Milwaukee, Detroit, Buffalo, Pittsburgh, Baltimore, and Philadelphia. The coal fields of eastern and western Pennsylvania also swallowed up hundreds of thousands of these people.

Although each of these groups has distinctive elements in its foodways, in the bustling give-and-take of immigrant societies in the industrial cities of nineteenth-century America, most of this distinctiveness was diluted. The Jewish immigrants were the most cohesive of these groups and were better able to maintain many of their foodways because they were associated directly or indirectly with religious taboos and traditions.

Most eastern and central European immigrants tended to live in socially or geographically isolated ethnic communities that allowed them to maintain their cultural integrity much longer than many other groups. They faced the greatest ethnic discrimination of any of the European immigrants as a group, and some, like the Slovaks, found it especially difficult to integrate with the earlier European immigrants. The church became a center of life in many of these communities, further contributing to the maintenance of old traditions. Most orthodox Christian churches are overtly ethnic—Ukrainian Orthodox, Carpatho-Russian, Russian; the Catholic churches serving the ethnic communities too took on a strong ethnic character. Polish Catholic churches were often fairly close to Irish and Italian Catholic churches but maintained a great social distance from them. Church services in most orthodox churches continued to be little Americanized well past World War II. Even then the transition was slow; services were typically partially or totally held in the native language of the parishioners. The continued use of the Julian calendar in the orthodox Catholic church dictated the celebration of religious holidays at different dates than their Catholic neighbors, further isolating them from other Christians. It was only in the post–World War II era that these residential ghettos began to disintegrate and strong Americanization began to take hold of everyday life, especially foodways.

The nineteenth-century eastern European diet tended to be very heavy by contemporary standards. The potato was a universal component of the daily and weekly dietary regime. Potatoes not only were consumed as side dishes but were also a common filler in many main dishes. Parading under a variety of ethnic appellations, what we call the piroghi* is essentially an eastern European ravioli; it can be filled with cheese, chicken, or other materials but most often it was filled with a mixture of mashed potatoes. Depending on the ethnic heritage of the cook, this comparatively heavy dish was then covered with sour cream, sautéed onions, or a variety of other sauces. A much tastier dish than it sounds, it is heavy by any standard.

Pork was the most popular meat, though fowl were also frequently served. Ducks and geese were consumed regularly in eastern Europe but were unavailable in many American meat markets and grocery stores. There is also a strong tradition of prepared meats, ranging from kolbasa and other sausages to what are generally perceived to be luncheon meats today.

Piroghi, stuffed cabbage, and stews (goulash, etc.) were some of the most common main dishes in these communities depending upon the specific eth-

*There is amazing continuity of dishes and recipes among the various groups that emigrated from Europe, but there is also wide variation in the spelling (and sometimes the names) for the same items. Consistency of terminology has been attempted, but note that the more common *piroghi* spelling was not used in the *Panis' Cookbook*, rather the uncommon *pirohi* spelling. Six additional names for this same dish were found in just two "local" cookbooks from the Midwest.

Pirohi

3 c. flour
2 eggs
2 T. butter
3/4 c. warm milk
3 T. sour cream

Mix ingredients and knead into soft, workable dough. Let rise for 10 minutes, covered with a warm bowl. Divide the dough in half and roll thin. Cut circles with large biscuit cutter, or cut in 4 inch squares. Place a tsp. filling in center of each, take one end and put it across another to seal. Pinch all around sealing well.

Drop pirohi into salted boiling water. Cook gently for about 5 minutes or until pirohi rise to top. Lift out of water carefully with slotted spoon and serve with sautéed butter onions on top.

Fillings:

Potato: Potatoes, American cheese, sautéed onion, salt and pepper.

Potato and Cottage Cheese: mashed potatoes, dry cottage cheese, sautéed onion, salt and pepper.

Prune: cooked, pitted prunes, sugar, lemon juice.

Sauerkraut: canned sauerkraut browned with onion.

Cabbage: chopped cabbage browned with onion, salt, and butter.

(*Panis' Cookbook*, 1978)

nicity of the family in question. These typically would form the bulk of the regular weekly meal rotation. An examination of church-sponsored cookbooks suggests that significant cultural interchange took place between these eastern and southern European groups. Such cookbooks carry all of the important national specialties of the sponsoring groups, but they also contain recipes for spaghetti and other dishes with broad appeal.

Ethnic assimilation among these groups was especially frequent in the smaller mining and mill towns, where there was a mixture of eastern and southern European residents. Ethnic intermarriage was more common because these small communities rarely had more than a single school. One may teach ethnic bias in the home, but somehow out in the schoolyard under the spell of a dark-eyed, mysterious Italian girl or a handsome, muscular, blond Polish boy, it is all forgotten.

Little of the general eastern European diet has been integrated into the American mainstream. Most Americans are aware of borscht and stuffed cabbage; most have never heard of piroghi and kolbasa. Few have eaten any of those foods. Goulash is a common term in American cooking, but most of the stews produced under this appellation have only a nodding acquaintance to the original. There is no real reason for this lack of acceptance, although the general trend toward lighter foods about the time that these foods were starting to have the potential to enter the dietary mainstream was certainly an important consideration.

Considering their numbers, Jews have long played an inordinately large role in the evolution of American cuisine. Twenty-three Sephardic Jews from Recife, Brazil, settled in New Amsterdam (New York) to start the long American tradition in 1654. Banned from many areas because of their religion, most early Jews in America were concentrated in a handful of comparatively open cities including New York, Newport (Rhode Island), Philadelphia, Charleston, and Savannah. It is believed that at least a portion of the French Huguenots that settled in Charleston and Savannah were also actually Sephardic Jews who had become Marranos (secret followers) in Provence before emigrating to America.

The vast majority of American Jews came in two nineteenth-century immigration waves. The first wave began about 1830 and was primarily of immigrants from Germany. This group tended to be relatively well educated and often had the resources to establish businesses in this country. Virtually all remained in larger cities, though many were peddlers covering large territories throughout the eastern United States. Many of these people eventually founded retail shops in one of their regular stops, where they were often isolated from the mainstream of Jewish life. The second wave after 1880 was primarily immigrants from eastern Europe. These settlers tended to have few financial resources and were more likely to work in the trades, especially in New York and other large northeastern cities.

For the most part, the Jewish populations ate much the same foods already described for the eastern Europeans or the Germans, with some exceptions.* Pork consumption was prohibited; only meat from cud-chewing animals with cloven hooves was allowed. Other food laws prohibited some combinations present in the traditional eastern European diet. Otherwise, there were many similarities in the cuisine of the Ashkanazic Jews and the eastern Europeans. Indeed, many purely eastern European dishes are perceived to be

---

*Sephardic Jews compose a very small proportion of the American Jewish population today. Their diet was largely a kosher Spanish-Portuguese menu. Some Charleston (South Carolina) food historians believe that they played a crucial role in the introduction of the now classic Carolina pilau.

Jewish; pastrami, corned beef, and thin-sliced, rare roast beef are perennial favorites for sandwiches in Jewish delis.

Bagels and cream cheese are probably the most visible contributions of Jewish dietary fare to the mainstream American diet. The origins of the bagel are lost, but it evolved in eastern Europe sometime before 1610, when the term was first used in print. Bagel bakeries began appearing in New York's lower east side in the mid-nineteenth century with both store sales and street hawkers selling their goods speared on long sticks as they walked the neighborhoods. Bagels remained popular until the 1920s, when they began to decline in popularity as the first generation of immigrants began dying out. Later generations typically perceived bagels as a Sunday morning breakfast treat. Bagel magnate Murray Lender noted, "Even up to the 1950s, you literally could not give a bagel away from Monday to Saturday" (Nathan, 1994, 85). This changed after 1951, when a new Broadway comedy, *Bagels and Yox*, brought national attention to the bagel, which in turn laid the groundwork for an article in *Family Circle* magazine using bagels as hors d'oeuvres and a recipe on how to create them in areas where they were not available.

Bagel bakeries were few and far between when Murray Lender returned to his father's bakery after his military service in 1955. He experimented with onion, egg, and pumpernickel flavorings to make the original bland product more interesting to a larger audience, and he convinced area supermarkets to begin carrying his products. In 1962 Lender created an automated bagel maker to allow the rapid expansion of production while creating a frozen product to allow the expansion of his sales territory. Kraft Foods purchased the company in 1984 and put its marketing muscle behind creating a national distribution. Frozen bagels were second only to orange juice in supermarket frozen-food sales until the recent rise of bagel shops in areas where bagels were unavailable in the past.

The bagel-sandwich shop began appearing as a competitor in the quick-service restaurant field in the late 1980s. The American breakfast, high in cholesterol and fat, is under attack by nutritionists, but there hasn't been much to replace it. Bagels and coffee offer an attractive alternative breakfast or lunch, and bagel shops are rapidly expanding, seemingly into every community in the nation. Riding the larger coffeehouse boom, the early entrants have been spectacularly successful. Bruegger's Bagels, the largest of the chains, had more than 200 stores in 1996, when it was acquired by a large multiconcept restaurant conglomerate.

Cream cheese was invented in upstate New York in 1872 by a gentile dairyman. Small amounts of cream cheese were manufactured and marketed

Matzos Pancakes

8 eggs, beaten separately	*1/2 c. matzos flour*
Sugar to taste	*1/2 t. salt*
1 lemon (juice and rind)	*6 grated cold boiled potatoes*

Mix batter evenly and lastly add the beaten whites of the eggs. Fry in small cakes in hot goose fat or butter. Serve with stewed prunes.
(Kander and Schoenfeld, 1903)

in the Northeast, but it did not receive particular notice until the Empire Cheese Company began production and distribution of the product as Philadelphia Brand cream cheese in 1880. The new product was quite popular in eastern European Jewish neighborhoods because it was similar to a cheese product found in eastern Europe. The Breakstone brothers opened their first dairy store in 1882 and soon also added Breakstone's Downsville cream cheese to their product line. The Breakstones heavily advertised their product in the New York Jewish community as the perfect complement to lox, matzos, and later bagels. Their efforts brought attention to cream cheese outside of the ethnic marketplace, attracting the attention of Kraft, which purchased the Philadelphia Brand company in 1928 and the Breakstones' company the following year. Nationalization followed after the Kraft media people instructed their comedians to slip the words cream cheese into as many jokes as possible on the *Kraft Musical Review* radio broadcasts.

Cheesecake soon followed. There are two types of cheesecake typically served in the United States: The Jewish version, generally called New York cheesecake, has a smooth cream-cheese filling; the Italian version is made with either cottage or ricotta cheese and tends to be lighter. Cheesecake recipes appeared in most nineteenth-century American cookbooks, but it was not until the mainstream society discovered both cream cheese and the glories of lunch at a Jewish delicatessen that the dessert began to become more widely known. The Kraft Company promoted a new home recipe in 1947 that became immensely popular, but ultimately it was Charles Lubin, a Chicago baker, who brought the old/new dessert to its current popularity. Discovering that he could freeze his high-quality cheesecake products with little loss in quality, he adopted his daughter's name, Sara Lee, as his brand name and began distribution throughout the Chicago area. Cheesecake freezes so well that a national chain of restaurants, the Cheesecake Factory, creates all of its signature cheesecakes in a California factory and ships them frozen to its stores despite the implication of local baking in the company name.

Smoked salmon, originally from Scotland (lox) but more recently from Nova Scotia (Nova), is increasing in distribution and is the most likely next food from this tradition to join the American mainstream. On a localized level, Jewish fare has made a great impact on the cuisine of those areas where large numbers of Jews live—most notably New York, Chicago, Miami, and Philadelphia. Knishes, lox, bialys, and kugels are as common in these communities as are grits and barbecue in the South.

Italian immigration largely paralleled that of the eastern Europeans. Small numbers of skilled Italians and Italian artisans came to the United States before 1880, but most entering after 1880 were unskilled and non-English-speaking southern Italians with no money. They took whatever jobs they could find. Almost 97 percent of these immigrants entered America through the port of New York, where they were rounded up by labor contractors and sent to construction projects, textile mills, and occasionally mining operations as work gangs. Unlike the eastern European immigrants, the Italians rarely worked the steel mills or the packing houses. Because these two groups competed for the same kinds of jobs, they rarely settled in large numbers in the same communities.

The geography of the 6 million Americans of single Italian ancestry today reflects this early history. The bulk of all Italian Americans is found in New York City and Philadelphia and surrounding areas in a line following the route of the Erie Canal to Buffalo and in and around Pittsburgh and Chicago. Smaller numbers are found in some western mining areas and railroad division points (communities where trains change crews and work crews reside) in the western and north-central states. The early geography of Italian foods originally reflected this distribution, although today most common Italian specialties have been integrated into the national cuisine.

Italian restaurants began appearing in New York and Philadelphia in the 1890s. These Italian restaurants targeted their Italian neighbors, but apparently a sufficient number of people of other backgrounds dined in these trattorias and interest in Italian foods spread. Although it is difficult to determine when spaghetti with tomato sauce and other southern Italian dishes began entering the American home, most pre–World War II cookbooks did not include recipes for these dishes. Widespread consumption of southern Italian foods did not begin until after World War II, reputedly as a result of returning servicemen gaining an appreciation of the cuisine during the war.

Pasta was at the core of the traditional Italian American menu. Italy is the home to many different cuisines, and the long-held belief that tomato sauce

Pasta for Eight

Mound the kitchen table with flour high enough to provide sufficient pasta for those coming to dinner. Create hole in top of mound and add one egg for each person to be served. Mix the ingredients by hand, sprinkling the dough with water as needed, until the dough can be rolled into a ball. Separate into pieces for rolling. Roll the dough as thin as possible with a broom handle so that it covers the entire table. Take the edge of the dough and roll it up across the table into a cylinder. After all dough is prepared, cut thin slices off the end of the dough cylinders and hang the "string" from the broom handle to dry.

Place in boiling water until tender and serve.

(Audria family oral recipe, ca. 1950)

was the essential ingredient in all Italian dishes has finally been laid to rest, as well as the belief that all red sauces are essentially the same. Prepared meats of all kinds were a specialty of the Italian food industry, and many have become widely used in this country, as well as a variety of regional cheeses. As in other immigrant groups, Old World Italian recipes were passed orally from mother to daughter, allowing a great variety of individual recipes to continue for several generations. The Americanized versions of these dishes thus also vary widely.

Flat bread dough covered with a variety of garnishes and baked has long been a staple in southern Italian homes and was brought to America with the second wave of Italian immigrants. Modern pizza baked from a dough is difficult to properly prepare in a traditional oven; thus the establishment of the first pizzeria in New York, often attributed to Gennaro Lombardi in 1905, was an important event in the evolution of American pizza. Little pizza was consumed outside Italian neighborhoods prior to World War II, but afterward it became a popular family meal for mainstream Americans in Italian neighborhoods, in college towns, and in other "open" communities after 1950. Contemporary pizza has virtually lost its Italian connection. One finds cheeseless pizzas and pizzas topped with artichokes, pineapple, barbecued beef, tandoori chicken, and virtually any other sandwich ingredient. Pizza has become so popular in American society that it is often rated as the favorite food in school lunch programs.

Interestingly, the nation's major chains of pizza parlors did not originate in the core area of Italian residency. Pizza Hut, the nation's largest chain, began with a single store in Wichita, Kansas. Domino's Pizza began in Ann Arbor, Michigan, a few years later. Other large chains include Little Caesars of Yp-

Friends and strangers are welcomed to a Texas supermarket opening with fajitas, an imported dish now ubiquitous across America. (Richard Pillsbury)

silanti, Michigan; Godfather's Pizza, beginning in Omaha, Nebraska; and Mazzio's, which also began in Wichita. No major chain has its roots in the northeastern heartland of Italian settlement, though numerous independent stores and small chains thrive. The frozen-pizza market, again largely dominated by corporations headquartered in the Midwest, has been growing in recent years as the quality of the product has improved. It is interesting that the largest chain of Italian restaurants is based in Orlando, Florida.

Less than a million single-ancestry Greeks live in America; yet they have played an important role in the history of American cuisine. The twenty-five years following 1880 was the period of greatest Greek immigration. Usually unskilled and uneducated, most Greeks began as unskilled laborers; some later went into business for themselves. Operating and working in restaurants were especially popular occupations for these immigrants. Ultimately, a large percentage of all the "railway car" diners were owned or operated by Greeks, who made them emporiums of classic American foods. Greeks also often operated candy stores, soda fountains, and early hamburger and hot dog stands, all bastions of classic American cuisine. Most Americans, however, were reluctant to even try traditional Greek dishes, and Greek food did not make significant inroads into the American standard menu until the rise of international cuisines in the 1980s. Today feta cheese and other Greek ingredients are increasingly found in restaurants and in the homes of the more adventuresome, but with the exception of Caesar salad, most Americans see little Greek food.

Recent Immigration Patterns

The passage of a system of quotas restricting immigration from most nations outside the Western Hemisphere was passed in 1921. This restrictive act was followed in 1924 by the Johnson-Reed Act, which lowered the total quota of immigrants to 150,000 per year. The act severely limited the immigration of people from southern and eastern Europe and totally halted the immigration of Asians. These restrictive measures reduced immigration to below 100,000 per year until after World War II, with negative net international migration taking place during the first four years of the 1930s.

Some exceptions were made to these draconian regulations. The *bracero* program (1942) allowed Mexican agricultural workers to harvest wartime crops. The War Brides Act (1946) allowed the returning heroes to bring their foreign spouses to the United States, and the Displaced Persons' Act was passed in 1948 to allow the nation to accept some of the millions of people affected by the shifting borders and political alliances following World War II. Special rules continued to be passed to handle political refugees, but it was not until the Immigration and Nationality Act (1965) that things changed significantly. Asian immigration was allowed after that year, but immigration of people from the Western Hemisphere was limited to 120,000 per year. Annual immigration rates rose to more than 200,000 in the 1950s because of an influx of political refugees, then exploded to a total of 3.3 million immigrants in the 1960s, 4.5 million in the 1970s, and almost 6 million in the 1980s. More than 1 million legal aliens entered the United States in 1989 alone.

The new wave of immigration brought significant changes in the nation's ethnic composition. Latin American and Asian immigration swelled enormously, and the United States had significant numbers of Chinese, Korean, Southeast Asian, and West Indian residents for the first time. The large number of immigrants from Asia and Latin America made Los Angeles the largest port of entry for immigrants. New York remained a center for the declining numbers of immigrants from Europe and for most Africans coming to the United States. Miami became the preferred port of entry for people arriving from the Caribbean basin and South America.

The relative economic stagnation in Mexico in the 1950s and 1960s accentuated the differences in economic opportunity in the United States and Mexico, instigating a massive northward movement of emigrants. There had been Hispano residents in the American Southwest before there were Europeans in the eastern United States, but the new economic conditions brought thousands of Mexicans across the border, primarily to Texas and New Mexico. The *bracero* program brought tens of thousands of new Mexican workers into California and Arizona for the first time after the 1940s. Many liked what they saw. All of these agricultural workers were by law supposed to return to Mexico, but some did not; many others returned legally and illegally to settle. The rush of Mexicans and other Hispanics from Central America after 1970 was focused on the economically booming California, not on Texas. Today there are more Americans of Mexican descent in California than in any other state.

Today virtually every corner of the United States has at least a few permanent Mexican American residents and often many temporary residents. Whereas California remains the most popular destination for workers coming across the border, every large city has a growing population of legal and illegal residents born in Mexico. Labor-intensive industries such as apparel, textiles, and food processing are most often the targets of these movements, but other work is also opening up in most service industries and in construction. Residential and commercial builders can realize enormous savings by hiring these workers. There is also a demand for workers who have skills that have often been lost in the United States.

The American Southwest was a part of Mexico until 1848; Hispano foods were obviously the only foods of that region until its incorporation into the United States. These foods continued to dominate those areas of early Hispano settlement even after American occupation and settlement expansion. The Hispano population of the Southwest maintained a strong independent identity from the incoming Anglo population, and its foods remained quite distinct, although they did begin diverging from those of northern Mexico. As many as six distinct Mexican American cuisines have evolved in the United States over the past century; the most well known of these is in the upper Rio Grande valley. Dan Arreola, a geographer studying Mexican American settlement throughout the West, persuasively argues that there are three zones of distinct lifestyles in southern Texas alone in addition to those in the upper Rio Grande, Arizona, and California.

Mexican food was accepted very slowly in the remainder of the United States. Anglos from the Southwest have always eaten a bit of Mexican food, but not until the late 1950s and 1960s did Mexican foods begin finding their way onto the plates of visitors to the region and residents of nearby states. True Mexican specialties first entered the mainstream through restaurants. Traditional American diners tended to pass over these selections, however, and concentrate on the more familiar sounding Tex-Mex specialties—ultimately driving most of the authentic foods from the menus.

Many Americans were first introduced to "Mexican" food by way of taco stands or street vendors. Glenn Bell's development of a chain of taco stands designed to be competitive with other fast-food outlets in the early 1960s did much to spread familiarity with the foods. His reddish-tan pseudomission concrete-block structures complete with plastic saguaro cactus garbage cans, the signature bell in the peak of the false front, and a gas campfire were unforgettable. Taco Bell has become a part of the Pepsi generation and gone national, competing toe-to-toe with the hamburger chains with varying success. The vast majority of southwestern, Mexican, or even more specialized restaurants ultimately are still forced to include the basic Tex-Mex American interpretations along with regional or local specialties.

Mexican food has made fewer inroads into the American home dining room. Nachos have become a common party snack. Sales of both flour and corn tortillas, as well as the hard shells used for tacos, continue to increase, but one rarely finds much exploration beyond the basic fast foods. Various hot peppers and salsas have been of increasing importance among those searching for new taste treats, but these ingredients are often utilized in nonstandard ways. Spanish rice was once a popular mainstream dish but seems to be less important today. Sales of frozen burritos and other foods for home consumption are also increasing, but Mexican food made from scratch is infrequently served in most American homes.

Other Latin American foods have made few inroads in the mainstream American diet. Caribbean restaurants are found in most larger cities, and jerked chicken is on the verge of developing a national following; but supermarket sales of primary ingredients are still largely restricted to stores serving Latin American neighborhoods. Black beans have found favor among many chefs creating lighter "Latino" foods, and they are also increasingly popular in soups and used in a variety of ways that are becoming common in areas with large communities of Cubans. Millions of Puerto Ricans have immigrated to the mainland over the past forty years, but almost all have remained concentrated in the Northeast, especially New York City. Few Puerto Rican specialties have spread beyond their ethnic neighborhoods.

Chinese restaurants are one of the oldest ethnic restaurant concepts in American restaurant history. The first Chinese to immigrate in numbers were brought as laborers to California and the West in the 1860s to replace expensive and often recalcitrant European-origin workers. More than 10,000 Chinese workers helped build the Central Pacific Railroad; others helped drain the Sacramento and Salinas River deltas and build diversion canals and other major projects. Some workers quit the large-project work gangs and went into mining or started businesses, most notably laundries and restaurants. Anti-Asian sentiments began to surface in the 1870s as the western economy slowed and European-origin workers resented Asians' willingness to work hard for low wages. The Anti-Asiatic League in California tried to ban the Chinese from jobs, housing, and schooling. Riots and attacks broke out in San Francisco and other communities that were aimed at forcing the Chinese living there to leave. Congress passed the Chinese Exclusion Act in 1882, prohibiting the immigration of male Chinese laborers and extending the ineligibility of Chinese for naturalization. Attacks on the Chinese continued in the West, and slowly the Chinese population retreated to the largest western cities, especially San Francisco, Sacramento, and Portland.

Virtually all of the early Chinese in America were from a small area on the lower Pearl River. Cantonese cuisine became identified in the American mind as the only Chinese cuisine. Chinese restaurants continued in the large urban Chinatowns and occasionally in smaller western towns, where small numbers of Chinese remained as miners and railway workers. One of the quixotic treats of driving in the intermontane West of Nevada and environs after World War II was to stop for a meal in one of the small, isolated communities in Nevada and discover that one of the four restaurants in town served Chinese food. A second often served Basque specialties.

The changing immigration laws of the 1960s coupled with the problems of the People's Republic and Hong Kong brought thousands of Chinese into America. A million or so Chinese currently live in the United States with roots in all of the major cuisine traditions of the culture. The vast majority continue to live in larger cities, especially on the West Coast, although a venerable Chinatown has been a part of New York since the nineteenth century. These new Americans have aggressively moved into many walks of life, though restaurants continue to be a popular entry-level business for those interested in retailing. The hold of Cantonese cooking on the American image of Chinese food has been broken, and it is possible to find restaurants serving at least the more important classic cuisines in most cities. Interestingly, however, there have been no successful restaurant chains specializing in Chinese cuisine, although there has been at least one well-financed failure.

Chinese food has made few inroads into the traditional American kitchen except for the widespread adoption of stir-fry cooking. Much of what is stir-fried, however, is not Chinese; nor do most Americans venture far from long-grain white rice in their cooking. This is not to suggest that Chinese food is not occasionally cooked by non-Chinese–origin Americans.

The end of the Vietnam War brought millions of Vietnamese and other Southeast Asian peoples into the United States. Large communities of Vietnamese, Cambodians, Laotians, and others have developed as these refugees have scattered throughout the United States. Despite aggressive plans by the federal government to spread these people as widely as possible, there has been a general trend toward concentration in a few eastern cities and the West. The original core areas in Los Angeles and the San Francisco Bay area have begun to spread and are the largest in the nation today. Communities of Southeast Asian American agricultural workers are increasingly found in the agricultural service communities of California and surrounding states and ultimately these people will spread into the small towns of the West, much like the Mexicans who preceded them. Small communities of Southeast Asian fishermen have also begun to develop from Texas to the Carolinas.

Vietnamese and other Southeast Asian ethnic restaurants have become a fixture of American ethnic neighborhoods, but there have been few inroads into home kitchens. Vietnamese bakeries are especially popular where they have appeared because of their interesting mixture of French and Asian influences.

Some Final Thoughts

Myriad ethnic groups have emigrated to the United States over the past 300 years. Most of the complexity and variety of their food heritages have been lost, the voracious American culture shredding the traditions of its new residents as it adopts some and rejects the rest. The initial western European food traditions set a pattern that still dominates the American national cuisine. New foods have been added to the culinary soup, but most often traditional ethnic favorites reach popularity only after significant modification to accommodate the same set of taste buds that created the McDonald's hamburger and Franco-American Spaghetti-O's. The foods that emerge seem Mexican or Italian or Chinese, but clearly they are American. The burrito, the pizza, french fries, and hoppin' John all have origins in some other nation but today are as American as the hot dog.

On the other side, the adoption of "American" foods by new immigrants seems to be taking place at a more rapid pace than in the past. An immigrant family of a century ago may have dined little differently for as much as three or more generations after its departure from the homeland. A recent series of studies of Hispanic nutrition by researchers at Colgate University suggests that although the generation that migrated may continue eating a fairly traditional diet, the children consume far less of their parents' ethnic fare than one might expect. The Colgate study's examination of even toddler diets found that they contained few tortillas, frijoles, and similar foods; American breakfast cereals, pizza, and sandwiches were far more typical. A Chinese colleague recently sent his preschool son home to Beijing with his wife to visit with the grandparents for a few months. The child dutifully ate the purely Chinese menu a few days before he began demanding to be taken to Pizza Hut for real food.

The arrival of almost 60 million immigrants over 200 years obviously has reshaped the American national diet. Less obvious is the impact of the settlement patterns of the newly arriving immigrants. German, eastern European, Vietnamese, and other ethnic groups emigrating to the United States did not settle randomly in the new land; rather they tended to concentrate in a few places. Their impact on the American diet has tended to be more highly concentrated in some areas than others. Ultimately, this has created food regions that are unlike our traditional concept of American regional cultures but are just as distinct. Terms such as New England, Middle Atlantic, and southern are increasingly meaningless when talking about contemporary American food regions. Certainly newly arriving immigrants are shaped by the traditions of their new homes, but they also shape these traditions. How many of the peculiar characteristics of nouvelle California cuisine stem from the large Asian presence there? Would there have been a southern cuisine as we know it without the African contributions? The immigrant has been a part of the evolution of American cuisine from the arrival of the first foragers moving southward along the Pacific Coast, but the immigrant's influence did not magically stop a hundred, fifty, or even twenty years ago. The shaping process continues and will continue as long as there are immigrants coming to this country.

8

Eating on the Town: Restaurants and the Diet

We were served a breakfast of beefsteak, sausages, stewed veal, fried ham, eggs, coffee and tea. . . . The next day we buried the innkeeper.

—Francis Bailey, 1754 (Bruce and Crawford, 1995)

The restaurant has become one of the most potent agents of change in altering the character of American dining. Virtually all of the new cuisines, cooking styles, and foods of the past several decades first entered the American diet in restaurant settings. Understanding the contemporary American cuisine scene thus requires understanding the restaurant industry, its history, and its current characteristics, especially now that more than half of every food dollar is spent in these establishments.

Body Food, Soul Food

Classifications are used to help bring order to chaos. Traditionally restaurants have been classified by the foods that they serve—Italian, seafood, steak—in the belief that the food served was the basis of a consumer decision process. The recent explosion in restaurant cuisines, paralleled by increasing consumer ennui, suggests that the type of food served has less and less to do with that selection process. A recent survey found that 70 percent of all diners had not selected their destination when they left their homes to dine out. As there generally are not large numbers of restaurants serving the same cuisine within the typical acceptable travel distance, it is clear that some other elements must play a role in the selection process.

An alternative approach to restaurant classification would be to categorize them on the basis of what they offer the consumer. Why does the diner choose a particular restaurant? Is the diner primarily going to a restaurant to satisfy hunger—in essence to feed the body—or is the consumer dining away from home to enhance or create an experience—to feed the soul? This is a complex question in that what constitutes body food to one person may be soul food to another. The concept thus is not about price or type of cuisine but about the experience and what it means to the diner.

"Body-food" restaurants, those that attract customers whose primary intent is to fill the stomach, are the oldest and most common form of restaurant. McDonald's is a classic body-food emporium. The goal of the world's largest restaurant chain is to serve food that nourishes the body with few frills or thrills. This goal is achieved in clean, efficient surroundings. The fare is purposely bland and nonoffensive to increase the customer base. Few

Dinner at Diamonds: Body Food, Soul Food

The journey after the call from the nursing home was a mad dash. We rushed to the Atlanta airport and then were picked up by Russ, my middle son, in the bitter cold of a December night at the airport in Philadelphia. We rode through the icy night without heat in his worn-out car; ice formed on the inside of the windows, cars littered the edges of the icy roads, and I had an icy feeling in my heart that we would not make it before my mother died. But after wrong turns, frantic calls, and near misses we got there, and her blue eyes sparkled in recognition from behind that mask of a body already lost. I spent the night trying to remind her of a past that now eluded her. Then in the morning she was gone, and I allowed myself to sleep for a few hours.

Richard insisted that we have a real Italian dinner for old times' sake before returning home that night, and we spent two hours searching for a place that was barely ten minutes from where we had started. Unsure, I had gone along with the idea and now was awash in the past I had left behind when I moved away so many years ago. Everyone had heard of the restaurant in Trenton, and everyone had directions that never seemed to get us there. Asking directions and riding down the worn, pocked streets had been frustrating. It had been a trip into the past. We passed the train station where I caught the train home from college on weekends. I pointed out to Russ, whom I had sheltered from my past, the tired building I had lived in during my first year at college. Finally, a clerk in a liquor store gave directions to the restaurant, now only two blocks away but down a narrow, unlikely street of row houses. Two young men in parkas bundled against the bitter cold told us that we had arrived and took over the car to send it to a parking lot or maybe a chop shop. We didn't care. Russell needed a car with a heater.

Entering Diamonds was like stepping back into the past. The foyer was crowded, and the maître d' asked if we had a reservation, knowing that we didn't. He found us a table away from the regulars in the very back. The place seemed tumultuous: too many tables crowded into the small first-floor rooms and a cacophony of voices celebrating the holiday season. Delicious smells from the kitchen—aromas we would never encounter in Atlanta no matter how hard restaurateurs there might try. An army of wait staff slipping between the tables, joking with the regulars, and being solicitous to the rest of us created a sense of déjà vu. The meal had been a good idea after all.

The red wine hit me like a rock. It brought memories of good times, and the tears came so hard that the wait staff began to worry. But Richard shushed them off and the past, shut out for so many years, flowed through me. The tomato aromas brought memories of hot summers of picking tomatoes and string beans and corn and canning them day after day.

(continues)

(continued)

"Patty Ann, hurry with those tomatoes," she called as I kept the handle churning the parboiled tomatoes through the peeling-seeding machine.

"Patty Ann, we need more corn. See if you can find another bushel back toward the creek."

"Patty Ann, . . . "

I had not thought about those summers of canning vegetables for a long time: the cool crisp mornings in the garden and gathering the growing harvest, the sweet smell of cooking tomatoes the first week, the acid smell of cooking tomatoes for the following weeks as we prepared shelf after shelf of tomato purée for sauce the next winter. It took a lot of tomato purée to feed five children pasta three nights a week.

My reverie was halted by the arrival of a platter of crisp bruschetta as an appetizer. Russ needed to get something in his stomach after two days of no food, little sleep, and much worry. The bruschetta (rather like a french bread pizza in today's parlance) was done to a turn, and looking at Russ, I realized that I had never talked much about growing up.

"We often had bruschetta when I was growing up on the farm, especially when my cousins came down for a Sunday visit in the summer," I began and then fell into thoughts about those times when I played all day in the fields and barn with my cousins from Newark.

Those were simple days. The cars would begin arriving about noon and the men dressed in their black suits, white, stiff shirts, and colorful ties would pile out; the women would follow with the children. We would gather in those first few minutes; we children, who saw each other only occasionally, would get reacquainted and eventually wander away from the adults to find fun without supervision. Playing down by the creek, we could see the men with black hair and black mustaches take off their black coats and smoke cigars under that great oak, as big around as an automobile, that protected the side of the house. They would squat and talk and smoke and drink a little wine while they waved their hands and discussed important things. We knew they talked of important things because they would switch to Italian whenever we came near enough to hear. We would then move further away, sometimes to the kitchen, where my mother and aunts would spend the day making fresh pasta for a ravioli dinner in the evening. Wearing their flowered housedresses, they mixed and patted and cut and chatted about children and homes, slowing only to shush us whenever we were too noisy. Marcy, my cousin, often spent the whole day in the kitchen, skittering under the table if there was a load noise, not realizing that this was not Newark. My cousins and I just laughed and giggled and never thought about adult things or why we lived down here on the farm and our aunts and uncles didn't.

(continues)

(continued)

"What happened to them?" Russ asked, breaking back into my reverie for a moment.

"I guess they are all still up in Newark; most of them were electricians and went into business with their fathers and got married. We just kind of drifted apart after I started moving around the country."

"Drifted apart" didn't quite capture what had happened. Actually, I had run as fast as I could and never looked back. Those had been good times, however, gathered around the great kitchen table in the evening when the food was finally ready—eating ravioli and talking, eating garlic bread and laughing, drinking a little vino and eating salad fresh from the garden and watching the whole scene in wonderment.

Suddenly I realized that dinner was over and I had barely noticed the spaghetti and meatballs go from my plate to my mouth. Russ was insisting that we should order the "real" canolli, which he had seen passing the table during the meal. Good canolli is so hard to find these days. That was always one of the great treats of those days. Uncle Mike used to go by Rosa's bakery on his way down to the farm and pick them up, and they would be the treat of those summer meals. But Russ would never know that; he would just think they were the perfect Italian dessert.

That sweet, creamy canolli was made from ricotta, not cream cheese, and had a crust so flaky that it melted, unlike those frozen sticks made in a factory somewhere that break into chunks of inedible cardboard.

The dinner was done. My tears were gone. We wound our way through the crowd of people who had gathered to wait for their evening meals while I was reminiscing. It was time to go home now, back to people who knew little about bruschetta or canolli or growing up on a farm in South Jersey a million miles from Newark. (by Patricia Pillsbury)

of us eat at McDonald's because we actually savor the flavors. We go there because the food is economical, it is served quickly, the coffee is always fresh, and the bathrooms are always clean.

The polar extreme of the body-food emporiums are the restaurants that draw customers primarily concerned with having a dining experience. Customers patronize these places to feed their souls (Pillsbury, 1990). The patrons of the Le Cirque, Spago, and Fog City Diner do not go to these places because they are economical or fast; indeed, patrons would be upset if they were hustled out before they were ready to leave. Certainly the coffee is always fresh and the bathrooms are always clean, but these places have become internationally famous because they provide their patrons with a dining experience that transcends the mere consumption of nourishment. In most cases these restaurants are only a stage set. One partakes of their pleasures to impress a client or to create a mood for an evening with a special friend or to celebrate a happy event. One might select an exotic restaurant to explore new food concepts or, conversely, go to the Fog City Diner to enjoy the foods of simpler days prepared in ways one's mother never dreamed. The dining experience thus is primarily about massaging the mind and creating an experience, not providing nourishment.

What has this got to do with understanding the American scene? Once we accept this dichotomy of restaurant cuisines it is easier to accept the dichotomy of cuisine generally. Increasingly, American restaurant dining habits demonstrate two mutually exclusive trends. One very successful group of restaurants is dedicated to maintaining the status quo. Nothing on the menu stands out in terms of flavor exploration; those innovations that do appear are incremental. These bastions of the status quo are by no means all quick-service emporiums. Comfort food comes in all kinds of guises, and more than 100 chains of midpriced, table-service restaurants have evolved over the past several decades. Their success is based on the maintenance of food quality, low cost, and hundreds of locations with exactly the same menus *and* food. Virtually all of this identical food is prepared in a commissary or factory (and indeed, increasingly the same factories) and the dishes are little more than heated on site. The food itself is immaterial; the consumer goes to Applebee's or Friday's or Denny's or any of the other midpriced table-service chains because they *are* all alike and the meal is predictable. Predictability is what body food is all about. Indeed, many fine-dining restaurants use the same approach to their fare; it is the decor, location, or service that actually provides a different dining experience.

Soul food has more to do with the dining experience than with the quality of the food. Many soul-food locations serve marginal food but evoke memories by creating or recreating images. Only a handful of restaurants actually provide truly innovative food—the conservative American attitude about new foods would not support many more of these places. And strangely, the very breadth of the variety of restaurants and cuisines has meant that the food itself is less and less significant in the dining experience. In the typical city, even the consumer who desires a particular cuisine may have twenty choices within commuting range. Ultimately, the restaurant selected may have less to do with the quality of the food than with the total experience. Indeed, one's *favorite* Italian restaurant may change depending on one's goal on any particular evening.

The Historical Evolution of the American Restaurant

The restaurant was an eighteenth-century invention. The first restaurant was French, of course, and the idea was rather slow to catch on in the American scene—of course. Americans did not eat away from their homes much in the eighteenth century, and then almost always while they were traveling. The impact of the Industrial Revolution on American ways of eating has been discussed at length, but one element not discussed was the effect of the new residential patterns on food establishments. Increasingly concentrated workplaces inevitably meant that larger and larger numbers of workers had to travel beyond easy walking distances from their homes. Horse trolleys and other public transit could move these factory and office workers from home to work site, but the midday meal, traditionally the most important meal of the day at that time, had to be either carried or purchased near the place of employment. Most of those whose incomes allowed the extra expenditures chose to purchase at least part of their meals near work.

Early Sources of Purchased Meals

Street vendors selling sandwiches, boiled eggs, and other prepared foods were the most common source of food for much of the working populace. Other workers scurried to nearby boarding houses, where being even a few minutes late meant that there would be little food left. Restaurants also began appearing, but these were comparatively expensive and little frequented by working men. The less expensive saloons and taverns that dotted the dock and commerce areas of the growing cities were popular, but if one drank too much those meals were expensive as well. Some were lucky enough to be able to frequent the numerous oyster houses found in the port cities, but most had to be content to get their midday meal either from a lunch pail or from a street vendor.

Seemingly an innocuous shift, millions of Americans gradually revised their attitudes about the foods and role of the midday meal. Buying prepared food away from their homes also forced them to begin sampling food prepared by strangers or, even worse, foreigners. Strangers or foreigners, these providers might prepare foods of the same name but often had their own ideas about the ingredients and preparations. Ultimately the street vendors gave way to vendors with horse carts in the industrializing cities of southern

New England and the Middle Atlantic, and those gave way to restaurants targeted at workers with low incomes and short lunch breaks.

Typically, the food served in the restaurants of the nineteenth century was not much different from what most patrons were consuming at home. Meats and breads were the bulk of these repasts. The menu of one of the earliest nineteenth-century restaurants, for example, was almost entirely composed of foods that were either fried or cooked in pots—soups, stews, and the like. Virtually all of the dishes were quite simple and consisted of a single ingredient—fried or stewed liver, fried fish, pork chops, fried eggs. These were largely the same dishes that the cookbooks implied were the primary fare at home. It is interesting that there is not a single fruit or vegetable on this particular menu, presumably reflecting the preferences of this previtamin era.

The rise of body-food restaurants was paralleled by the rise of soul-food emporiums, although their numbers were quite limited until the rapid expansion of the middle class after World War II. Prior to that time the nation's finest chefs worked for the very rich. Throughout the nineteenth century, most of the wealthy had some type of cook living in the household, and there was little demand for soul food in the open marketplace. There were fine restaurants in America and every city had a few, but just a few. Very often these establishments were located in the finest hostelries so that the guests would not have to go out on the town to find proper sustenance. A few important restaurants did evolve independently of hotels in the larger cities. Thomas Downing's Oyster House at 5 Broad Street was one of the first famous soul-food restaurants in New York outside of hotels and private clubs. Downing operated an elegant establishment that attracted bankers and financial men from Wall Street in large numbers; female patrons and their escorts ate in a separate dining room. It was one of the few public establishments at that time in New York in which a proper lady could dine.

The most famous fine-dining emporium of New York was Delmonico's. Although the first store opened in the 1820s, the restaurant began developing its well-deserved reputation in the 1830s. Moving northward with the advancing frontier of wealthy residential areas, the restaurant ultimately settled at 864 Forty-fifth Street. The restaurant was finally forced out of business in 1917 by changing times and prohibition.

This was a time when thin was not in. Calories, nutrition, and balanced diets were not yet considered, and the rich believed that sumptuous repasts were just one of the ways in which they demonstrated their superiority over the masses. Charles Ranhofer (1893), one of the restaurant's most famous chefs, immortalized his thirty-two-year reign over the restaurant's kitchen with a 1,183-page cookbook and treatise on Franco-American cooking. Virtually all of the menu items served were at least known to the average reader; the special-occasion menus seem impossible to have been created with the technology available at that time. Cost clearly was no object in the preparation of these special menus. Ranhofer's extensive menu compilations suggest that the restaurant had a different menu for every day of the year, and there were more than 100 ways of preparing eggs for breakfast alone. An August daily breakfast menu, for example, might include salmon quenelles, tenderloin of beef, green peas with braised lettuce, frog legs, roast chicken, an omelette, cheese, and fruits (Ranhofer, 1893). Luncheons, buffets, and dinners provided diners with even more complex and exquisite repasts and surely supported the common belief that a robust body was a healthy body.

Mass Production and the Restaurant

Standardization and mass production became the bywords of the late-nineteenth-century industrial age. Mass production of food made it possible for restaurant operators to confidently make long-range plans regarding their cuisine and printed menus. This mass-production mood shaped the restaurant business in three ways: It encouraged the development of the technology to allow large numbers of stores that all looked exactly alike; it created the impetus for aggressive restaurateurs to emulate their other retail brethren and to expand into hundreds of locations throughout the nation; and it allowed the standardization of menus at those new locations.

A number of new restaurant types began appearing during this period. The diner was the first of the mass-produced restaurants. It epitomized the Industrial Revolution. Created by a poor, uneducated entrepreneur in 1873, the diner was turned out by great factories in a process that owed nothing to the inventor. They were sold as turnkey operations complete with everything except the patrons and the food. The earliest diners were operated by the previous generation of street vendors who had been successful enough to purchase these grand food emporiums, but ultimately the vast majority of diner proprietors were Greek immigrants. Though Greek food rarely appeared in these places, the Greek love of good food, good coffee, and good desserts was evident.

One of the advantages and downfalls of diners was that each was operated by an independent agent. Thus consistency and standardization, which seem to be especially important to the American consumer, were lacking; hence if the same item was ordered in two locations, it tasted different. This was a serious problem. For example, diner hash browns along the southern New

The typical diner was cramped and served comparatively simple foods in Spartan surroundings. (Library of Congress)

England coast traditionally were potatoes fried with onions, peppers, and paprika. Hash browns a little further down the coast lacked the green peppers and paprika. Hash browns even further south were just fried potatoes. Today, of course, hash browns are likely to be deep fried slabs of grated potatoes.

It is all too easy to overemphasize the importance of diners in the restaurant revolution. There were thousands of diners concentrated in the Northeast, and there were tens of thousands of grills and cafés serving essentially the same fare in less romantic buildings elsewhere in the country. And many of these too were operated by Greeks. The passing of the diner reflects changing consumer demand. Consumers began to demand more from their eating places—cleaner facilities, more modern foods, and continuity of food quality. Whereas individual diners could provide these things, the diner as a class could not. Diners were individually owned and operated. Some were clean; some were not. Some offered good, memorable food; some offered just as memorable meals at the opposite end of the spectrum. In the end, the very similarity in architecture that made them almost instant icons also made them all too easily pigeonholed by the public as something to be avoided.

A second classic restaurant during this period adopted the factory mode on the inside rather than on its exterior. The cafeteria epitomized the ideals of the Industrial Revolution in its mass production of food, interchangeability of items on the food lines, and rapid handling of patrons. Invented in the 1880s, the concept received international notice during the Chicago World Exposition in 1893. Called honor houses because the customer told the cashier what was eaten on departure, this new self-service establishment reduced labor costs, increased efficiency, and placed commercial food preparation on the all-important production line. Cafeterias allowed for the presentation of hot foods at "factory" prices. The first major investor in the new concept was Bill Taylor, who virtually walked out of the Chicago Exposition and leased the site for his first store. His small beginning evolved into one of the largest cafeteria chains in the nation. Most cafeterias, however, were independently operated and had many of the same problems as diners in establishing credibility with the casual customer.

The Fred Harvey chain is generally given credit as the first restaurant chain in America, but in actuality there were a number of predecessors. Multiunit ownership became increasingly common after 1830 as restaurant owners purchased second and third stores. Fred Harvey contracted with the Atchison, Topeka, and Santa Fe railroads to provide food for passengers at railroad stops in 1876. His chain of railroad restaurants never exceeded forty at any one time, much in contrast to the hundreds of stores associated with the Waldorf and White House sandwich chains or the Binford, Childs, and other cafeteria chains cropping up in the eastern industrial cities at the same time. Harvey's menus by necessity were simple, as the orders were sent ahead to the restaurant by blasts of the locomotive whistle and the customers had to receive and consume their meals in little more than a half-hour.

The last of the major types of restaurants that evolved during this period were the sandwich shops. Most were initiated by individuals with little thought of expansion, but as success struck, some rapidly expanded to multiple locations. Operating with minimum investments and small trade areas, the sandwich chains were able to rapidly expand in the growing industrial cities. Their simple menus and low prices made them quite attractive to many blue-collar workers, and it was not uncommon for a chain like the Waldorf to have forty or more stores in a single city.

The ultimate expression of the new rule of automation, interchangeable pieces, and factory order was the White Castle chain and its imitators. The White Castle chain began in 1916 in Wichita, Kansas. Distinctive buildings were in vogue, and this chain created a squarish building sheathed in white, enameled metal with a crenellated facade to reflect the company name. Not obvious to the public but very much in the mind of the company was the fact that these prefabricated buildings could be disassembled just as quickly as they were assembled. The company could thus rent open space on a corner or along a major thoroughfare that was not immediately targeted for development. If the landowner decided to raise the rent or cancel the lease, the stores could simply be placed on a truck or disassembled and moved almost overnight, much like the movable diners of the previous generation.

The most important contribution to the restaurant scene was the heavy iron sheet that Walter Anderson had shaped into a grill much like those used in Belgium and the Netherlands. The adoption of the grill as the chain's primary cooking surface instantly determined the fare that could be provided. The grill allowed for the preparation of a large quantity of fried foods at one time, and it wasn't necessary to constantly clean cooking pans; it also defined what could not be cooked and served. Anderson, one of the claimants to the invention of the hamburger sandwich, also served hash browns, eggs, pancakes, and grilled sandwiches. He could not easily cook soups, stews, vegetables, and other dishes requiring a pan or pot. The later development of the deep-fat fryer added a second cooking mechanism to these restaurants, but during the early years the menu revolved around food that could easily be cooked on an iron grill.

White Castle, White Tower, Little Tavern, Krystal, and similar chains set the foundations of the new order of restaurants. Their streamlined menus got Americans used to selecting from a few items. Their new atmosphere opened the door to the introduction of new foods, and the innovators of the time created a whole new class of foods to be served in these emporiums. Indeed, all of the foods important to the fast-food genre were perfected over a span of about twenty or thirty years around the beginning of the twentieth century.

The Beginnings of the Fast-Food Genre

The industrial age spawned a new fare fit for the times. The pace of everyday life quickened as people were more likely to be away from home for more and more meals. Although virtually all early restaurants served foods that were the same as their patrons ate at home, many Americans had neither the time nor the money to patronize these places on a daily basis. These factory and blue-collar workers usually brought their own food from home, but rising incomes and increased discretionary spending made an occasional purchased lunch possible. The problems remained the same, however: The meals had to be inexpensive, they had to be able to be eaten quickly—often standing or on the move—and they had to be so simple that the seller could prepare them with minimal facilities. Finally, as always during these times, they had to be familiar. Though created individually, the foods that were reinvented and that gained popularity during this period have come to be called "fast food."

Fast food was the first class of foods that was created outside the home to be consumed outside the home. None of these foods would have been possible prior to the industrial era, though each has precursors stretching back to the dawn of modern food. All have ingredients that were reinvented during the industrialization of the food industry. In their purest forms, most of them also required the creation of new equipment for their preparation.

The sandwich was named after the Fourth Earl of Sandwich, a notorious gambler and wastrel who ordered bread and meat "sandwiches" during his marathon gambling binges. Though the term dates from 1762 and a recipe appeared in Leslie's 1837 cookbook, the sandwich in America dates from the rise of commercial bakeries and the appearance of presliced, soft wheat bread in grocery stores in the late nineteenth century.

The popularity of the sandwich increased with the appearance of chains of sandwich shops throughout the eastern seaboard in the 1880s. City workers found the new food tasty, inexpensive, and easy to eat in a hurry. Because the sandwich was easy to prepare, it was soon made frequently in the home, especially for lunches that were carried to work or school. Its late appearance in cookbooks suggests that it didn't receive much respect until the 1930s.

The hamburger is the most famous of the sandwiches, though purists contend that a proper hamburger must be made from a patty of cooked ground meat placed in a bun, not between two slices of bread. Ground meat became increasingly common around the beginning of the twentieth century as Americans ate more fresh beef. The growing popularity of sandwiches triggered the invention of this item, but the when and the where of its invention is in dispute and will probably never be known. The most frequently cited originator of the hamburger is Walter Anderson, founder of the White Castle chain. The fact that White Castle is one of the nation's larger restaurant chains probably helped support this assertion. It is quite possible that one of the other contenders actually invented this popular sandwich.

The hamburger initially was the kind of sandwich found at White Castles and diners and was less common in more expensive restaurants. It was rarely eaten at home until at least the 1930s. The *Good Housekeeping* cookbook of 1922, for example, has a recipe for a broiled "hamburg" steak, but no sand-

TABLE 8.1 Leading Restaurant Chains by Sales, 1995

Company	*Sales ($M)*	*#Units*	*Primary Product or Service*
1. McDonald's, Oak Brook, IL	29,914	18,380	hamburgers
2. Burger King, Miami, FL	8,400	8,030	hamburgers
3. Pizza Hut, Dallas, TX	7,900	12,140	pizza
4. KFC, Louisville, KY	7,725	9,668	chicken
5. Taco Bell, Irvine, CA	4,925	6,700	fast Mexican
6. Wendy's, Dublin, OH	4,500	4,700	hamburgers
7. Hardee's, Rocky Mount, NC	3,360	3,463	hamburgers
8. 7-Eleven, Dallas, TX	3,080	15,000	convenience store
9. Subway, Milford, CT	3,000	11,420	sandwiches
10. Domino's Pizza, Ann Arbor, MI	2,650	5,257	pizza

SOURCE: *Restaurants and Institutions*, July 1, 1996.

wich. Indeed the instructions for making the patty are so precise that the author must have believed that many readers would have little idea as to its look. Fannie Farmer's 1941 edition contains recipes for sandwiches, but no hamburger. The 1942 edition of *Joy of Cooking*, which focused on fast meals, does contain a recipe for broiled hamburger sandwiches. Hamburgers were popular in the 1920s and 1930s as restaurant fare but did not reach their current position in American cuisine until the 1950s and the rise of drive-in restaurants. Many of the earliest drive-ins during the 1930s and 1940s, for example, featured barbecue sandwiches and other "hand foods." Many restaurant forecasters in the 1950s and 1960s believed that the hamburger market was reaching saturation and that chicken and other foods would eventually dominate.

The hamburger today has taken on an aura of its own. There are hamburger societies, McDonald's groupies that visit and collect memorabilia, and even restorations of early hamburger palaces. The hamburger is the favorite food of a large percentage of all children and is eaten regularly by many Americans. The largest ready-to-eat food purveyor in the world is essentially a hamburger restaurant chain, and four of the top ten restaurant chains specialize in hamburgers (see Table 8.1).

Cooked sausages served on bread were a common item in the street vendor's basket. The hot dog had to wait for the creation of the modern wiener, first produced in quantity in the late nineteenth century. There are a variety of claimants for the term "hot dog"; the *Oxford English Dictionary* dates the first use of the term at 1900. Harry Stevens, the director of catering at New York City's Polo Grounds, is reputedly responsible for having the bun heated and condiments made available. His vendors roamed the stands calling out,

"Red hots! Get your red hots!" T. A. Dorgan, a Hearst sportswriter for the *New York Tribune*, first used the specific term "hot dog" in his column in 1903.

Strangely, the hot dog is one food that is served by the millions in homes but is still perceived to be best when eaten away from home. The wiener may be boiled, grilled, or fried before being placed in a bun depending on where it is served. Mustard is almost the only universal condiment, although ketchup has been increasing in use since World War II. Sauerkraut and coleslaw are popular additions in strongly Germanic areas; chili is also popular in many areas. Although chili is a southwestern food, Cincinnati has long had an independent chili tradition, especially in fast-food restaurants. Chili dogs seem to be most common in the Midwest. Other versions include the New York System, found in Rhode Island and created by cutting a long sausage into sections and covering the concoction with a meat sauce; the Kansas dog, covered with mustard and melted cheese; and the Chicago hot dog, which has a poppy-seed bun.

Oscar Mayer is the largest manufacturer of wieners in the world. Founded in 1873 in Madison, Wisconsin, by its namesake, the company today is a part of the Philip Morris food empire. Oscar Mayer was especially efficient in establishing a unique identity for its product through jingles—"I wish I were an Oscar Mayer wiener"—and, after 1936, the wienermobile. The 1936 model was thirteen feet long and advertised "German Style Wieners." Twenty-four of these unique automotive billboards have been built in five different designs. The most recent were built in 1995 after extensive wind-tunnel testing. These twenty-seven-foot fiberglass vehicles can reach speeds of ninety miles per hour. The company has ten roving three-person teams covering the United States and promoting its products. Many of the older vehicles continue to operate overseas, including three in Spain, two in Canada, and one each in Mexico, Japan, and Puerto Rico.

The concept of baking a thin sheet of dough covered with an assortment of toppings is at least 1,000 years old and possibly may date back yet another millennium. Something approximating pizza was being baked in the Naples area possibly as early as the eighteenth century. One possibly apocryphal tale states that tomatoes were first added when Raffaele Esposito baked a pizza for Queen Margherita and replicated the national colors: red (tomato sauce), white (mozzarella cheese), and green (basil). Pizza in America has evolved into a distinct food quite different in taste from the European product. The first pizzeria was opened in New York in 1906 (some date the first store as early as 1888), but this restaurant form was rarely seen outside of Italian

To Fry Sliced Potatoes

Peel large potatoes, slice them about a quarter of an inch thick . . . dry them in a clean cloth, and fry them in lard or drippings. Take care that your fat and frying pan are quite clean; put it on a quick fire, watch it, and as soon as the lard boils and is still, put in the slices of potatoes, and keep moving them till they are crisp; scoop them up and lay them to drain in a sieve; send them up (to the dining room) with very little salt sprinkled on them.
(Randolph, 1825)

neighborhoods in the nation's largest Italian concentrations until after World War II.

Pizza grew rapidly in popularity in the 1950s with the increasing importance of eating away from home. Today Americans consume more than twenty-three pounds of pizza per person with more than 600 million pizzas sold through home delivery alone. Pizza is especially popular as a finger food among teenagers and as an accompaniment to group activities. Pepperoni is the favorite topping among Americans, and anchovies are so rarely ordered that one has to wonder why they are a standard topping on virtually every menu.

There are several different versions of "American" pizza. The Midwest (especially Chicago) has long favored a pizza with a thicker crust than those found on the East Coast. More recently there has been a trend toward exotic toppings—pineapple chunks, artichoke hearts, and barbecued-chicken pieces.

The french fry, like many of the foods of this industrial age, has origins shrouded in mystery. The term seems to be American and stems from the method of slicing food into narrow strips, as in French-cut green beans. Recipes for fried potatoes appear in virtually every nineteenth-century cookbook. The transition from Randolph's "To Fry Sliced Potatoes" (1825) to Ranhofer's "Potatoes Fried and Channeled" (1893) is a very short one, and it is likely that others made the necessary transitional steps.

The concept of the french fry may date from the nineteenth century, but widespread consumption did not occur until the development of efficient deep-fat fryers in the 1920s. Most diners and grills served fried potatoes prior to that time, but true french fries did not become an important dietary item until the quick-service revolution. The McDonald brothers, for example, did not include french fries on their menu when they made the transition from a drive-in to a walk-up restaurant in 1949. It was only later that they recognized the importance of this high–profit margin side item in maintaining a strong balance sheet.

Soda water occurs naturally, and flavored carbonated beverages were available throughout the nineteenth century. The first artificially carbonated and flavored beverages appeared in the 1830s. Colas first appeared in 1881; Coca-Cola followed in 1886 and became a registered trademark in 1891. Coca-Cola's chief competitor, Pepsi-Cola, appeared in a New Bern, North Carolina, soda fountain seven years later with bottling beginning in 1904. Considering this origin, it is not surprising that southerners drink several times more cola than residents of any other region in the nation. The Northeast tends to prefer the older, more traditionally flavored soda waters including root beer, ginger ale, and cream soda, whereas the West tends to favor more "natural" soft drink flavors.

The explosion of soft drink, or soda water, consumption in the twentieth century has been one of the truly amazing transformations in American food. Rising from a minuscule consumption around the turn of the century, soft drinks have become one of the most important beverages in the United States today (see Figure 8.1). The value of soft drink production in the twenty years after World War II quadrupled from $748 million in 1947 to $3,173 million in 1967; the coincident consolidation process reduced the number of competitors by almost one-half. The Coca-Cola Company is the largest of the survivors with worldwide sales of $16 billion for all products. PepsiCo, the second largest cola manufacturer, had worldwide sales of $24 billion, but a significant part of that income includes several large non-soft-drink subsidiaries (see Table 5.3). Independent regional bottlers of soft drinks, such as Big Red in Texas and Oklahoma, have almost disappeared from the marketplace.

Restaurants Between the Wars

Automobile registrations jumped from about 2 million to over 8 million between 1915 and 1920, spurring the beginning of an entirely new way of life in America. Annual automobile sales passed 2 million per year in 1922. There was only one automobile per twelve households in 1920; that number changed to about one per household by 1930. Many households did not own automobiles at that time, of course, but the impact of the new transport medium was too pervasive to ignore. The first drive-in, the Pig Stand, was established on the Dallas–Fort Worth Highway in 1921. It featured barbecue (beef) sandwiches, and the ensuing chain was strongest in the greater Texas region, although it was bicoastal at one period.

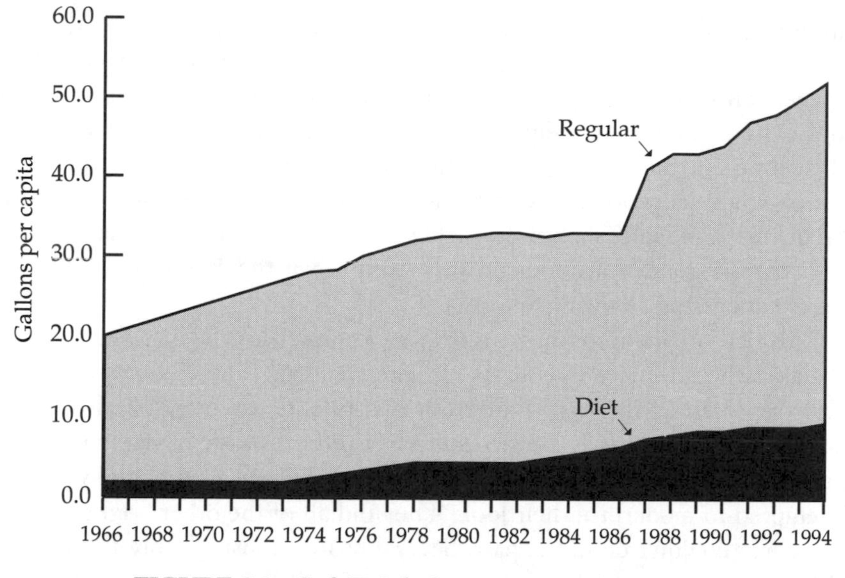

FIGURE 8.1 Soft Drink Consumption: 1966–1994

G. Kirby with his Pig Stand chain did not have the drive-in field to himself for long. A&W Root Beer was founded two years later in Lodi, California, with stores initially concentrated in central California. The concept of pulling up to a store and having a waiter serve a mug of ice cold root beer was too good to restrict to the Central Valley, and before long the chain had expanded throughout the arid West. One patron in Salt Lake City, Utah, liked the concept so much that he bought a franchise for Washington, D.C., and moved there with only limited resources. J. W. Marriott opened his first root beer stand on May 20, 1927; he later thanked Lucky Lindy for making his first day in business a success. People spent that entire day standing out in the streets, listening to news of the Lone Eagle's first flight to Paris, and buying frosty mugs of A&W Root Beer from his new store.

Drive-ins were most popular in the Sunbelt, where carhops could serve patrons most of the year, though drive-in restaurants eventually appeared in every state. The hamburger was not the dominant signature food of all drive-ins. Barbecue continued to be quite popular throughout the South, and hot dogs, root beer, and a variety of other foods became the signature dishes of drive-ins around the nation.

The McDonald brothers opened their first restaurant in Los Angeles in 1937. They moved to San Bernardino, California, in 1940. They became so frustrated with the constant problem of carhops not showing up for work or attracting young men who hung around the parking lot, making other patrons nervous, that they closed the store for a few months in 1949 and reopened with a redesigned building and business featuring "walk-up" service and a new simplified menu of fifteen-cent hamburgers, nineteen-cent cheeseburgers, milk, buttermilk, coffee, orangeade, root beer, cola, pies, and potato chips. Popcorn was sold from a vending machine outside the store. French fries and milk shakes were added to the menu a few years later. Lines of patrons extended around the store and down the street, and imitators started appearing all over California.

The rise of the automobile mortally wounded the White Castle–style chains, although most still survive in altered form. These chains of simple grills represented the past—inner-city apartments, streetcars, and belching factories—that the newly escaped suburban refugees were trying to leave behind. These companies built stores along the suburban arterial highways and attempted to modernize their look. They did all of the other "right" things. They just couldn't erase the past. Several chains, most notably Krystal and White Tower, began franchising territories from their new competitors and competing with themselves as Wendy's, Burger King, and other more modern quick-service franchise operations. Waffle House, the largest and most recent of the surviving white-box imitators, has succeeded largely by denying its origins. There are now 930 Waffle Houses scattered across the South, increasingly at freeway interchanges. White Castle remains as the seventy-fourth largest chain with 296 stores primarily in the Midwest and Northeast. Krystal, which thrives mostly because of the restaurant activities of its subsidiaries, is the eightieth largest chain with 336 units but has only about 90 percent of White Castle's sales.

Drive-ins were not the only beneficiaries of the automobile revolution. Suburban automobile-oriented general cafés also flourished. Howard Johnson's began selling ice cream, parlaying this early start into the first successful, large-scale, automobile-oriented general restaurant chain. This chain too fell victim to history but was replaced by a new breed of clones after World War II. The 1960s was almost an age of the suburban coffee shop with new chains opening all over the country. Denny's was the most prominent of these, though Perkins Pancake Houses, Sambos, Village Inn, and others were never far behind.

The food of these new automobile-oriented places was little different than that of their predecessors; it was just mass-produced. An extensive breakfast menu was served twenty-four hours a day, a range of sandwiches focused on the hamburger, and a variety of hot entrées for lunch or dinner included breaded, deep-fried shrimp; hot pork and beef sandwiches; meat loaf; and a

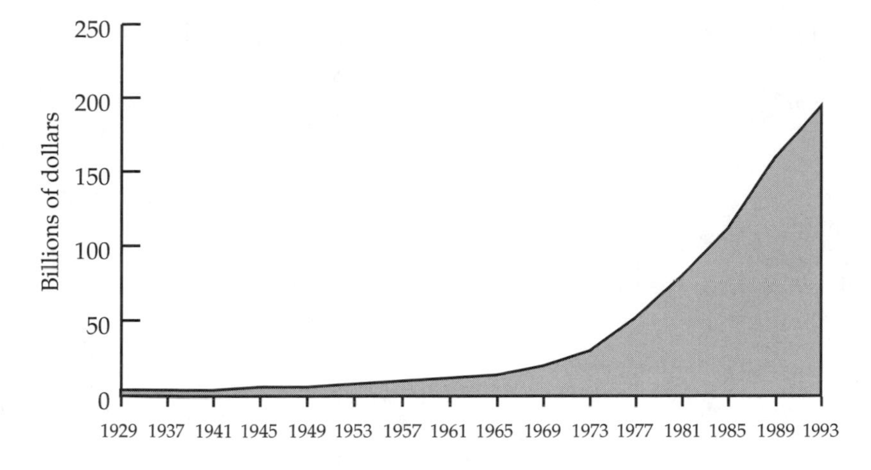

FIGURE 8.2 Restaurant Expenditures: 1929–1993

few other general dishes. There was nothing new here, just new glass and polished-metal buildings, bright interiors, and factory-created food.

Contemporary Restaurants and Contemporary Fare

The role of the restaurant industry has exploded over the past twenty years in the United States. There are almost 350,000 restaurants grossing $173 billion in sales in the United States today. The growth of the industry in the past twenty years is mind-boggling; sales have more than tripled since 1975 (see Figure 8.2). Growth has brought industrywide restructuring; most notably, chain restaurants play an increasingly prominent role in shaping the industry. The top 400 chains accounted for only $8.7 billion in sales in 1965, about 18 percent of all eating-away-from-home sales that year. Today the 400 leading chains account for 51.4 percent of sales, and the top 100 control 86.1 percent of chain sales. McDonald's alone had sales of just under $30 billion in 1995 (seeTable 8.1).

Thirteen cents out of every dollar spent in chain restaurants in 1995 was collected at a McDonald's cash register. McDonald's continues to set the pace for quick-service chains, and most do little more than emulate the leader. Pizzerias are the second most important chain-restaurant segment with about 10 percent of sales, followed by chain restaurants in lodging establishments, those devoted to family dining, those specializing in chicken entreés, and dinner houses.

TABLE 8.2 Market Segments of 400 Top Restaurant Chains, 1995

Segment	*Sales ($M)*	*Percent of Top 400*
Burgers	51,208	33.2
Pizza	15,594	10.1
Lodging	11,966	7.8
Family dining	11,751	7.6
Chicken	10,439	6.8
Dinner houses	8,193	5.3
Sandwiches	7,753	5.0
Mexican	7,240	4.7
Steak/barbecue	6,875	4.5
Seafood	3,881	2.5
Convenience stores	3,571	2.3
Italian	3,019	1.9
Cafeterias	2,306	1.5
Asian	513	0.3
Total	154,211	100

SOURCE: *Restaurants and Institutions*, July 1, 1996.

Though it is impossible to know what people are actually eating in restaurants, we can make some generalizations. The American love affair with the hamburger, which began in the early 1950s, continues unabated. Although American consumption of red meat in the home is dropping, total beef sales are supported by restaurant hamburger sales. The hamburger chains, of course, offer a variety of other foods including several different chicken sandwiches, but the hamburger continues to dominate sales with McDonald's, Burger King, and the others.

The basic pattern is changing. Boston Market, a chain featuring rotisserie chicken, grew 107 percent in 1995. Other fast-growing chains in the quick-service sector include Papa John's (pizza), Church's Chicken, and Carls Jr (a West Coast hamburger chain). KFC (Kentucky Fried Chicken) continues to be the third largest quick-service restaurant chain, followed by its stablemate in the PepsiCo restaurant group, Taco Bell. The hamburger may continue to dominate this segment of the industry, but its hold may be in jeopardy.

Chain restaurants reflect an interesting dichotomy in American culture in that they must appear to be new and fresh to be successful, yet they must serve traditionally prepared food for sales growth. Mainstream American food dominates this industry. Similarly, the most successful of the independent restaurants tend to walk a narrow line between tradition and innovation. Those independent restaurants with the greatest annual sales have comparatively staid menus and few cuisine innovations (see Table 8.3). This

TABLE 8.3 Leading Independent Restaurants, 1995

Name and Location	*1995 Sales ($M)*	*Patrons (000s)*
1. Tavern on the Green, New York, NY	30.5	575
2. Rainbow Room, New York, NY	30.0	472
3. Smith & Wollensky, New York, NY	20.4	395
4. Bob Chinn's Crab House, Wheeling, IL	20.3	914
5. Sparks Steakhouse, New York, NY	19.6	289
6. Joe's Stone Crab, Miami, FL	15.1	425
7. The Manor, W. Orange, NJ	14.0	275
8. Scoma's, San Francisco, CA	13.1	436
9. The "21" Club, New York, NY	13.1	152
10. Spenger's Fish Grotto, Berkeley, CA	11.9	910

SOURCE: *Restaurants and Institutions*, March 15, 1996.

is not to say that they are not very good. Certainly Scoma's is one of my favorite seafood restaurants in the country, but the menu is a monument to tradition, not the culinary innovation that characterizes the San Francisco Bay area. The goal of its owners is the same as that of the managers of McDonald's, to provide a quality, predictable, inexpensive meal. As they say on the wharf outside, "You get what you pay for."

Restaurants and Dietary Change

If neither chain nor independent restaurants provide innovation, how can it be said that the restaurant has been the center of culinary innovation in this country? Innovation is not a mainstream phenomenon; the most innovative restaurants are the smaller ones with young chefs who see cooking as an adventure and feel the need to depart from the well-trodden path. The problem is that opening a restaurant is an expensive undertaking. Even fairly simple restaurants like an Applebee's or a McDonald's require more than a $1 million investment. Almost by necessity, innovation takes place in smaller places with minimal accoutrements. Unfortunately, as the restaurant begins to receive some fame, it is unlikely that it will continue innovating; once the financial backers realize a return on their investment, they tend to become fearful of losing future rewards. Further innovations are incremental or not made at all. This is one of the factors that leads so many young chefs to periodically change location and ultimately to attempt to run their own stores. Unfortunately, few innovators are also good managers.

9

A Contemporary American Diet

Cooking in America is increasingly becoming just a hobby.

—Dennis Leonard, 1996

The American diet has changed more in the past forty years than in the previous 150. A portion of this change is due to technological and immigration shifts discussed previously, but a good portion also stems from the nation's restructured lifestyle after World War II. During a short period the central cities decayed, an entirely new suburban lifestyle evolved in neighborhoods built on what had been farmland, and the very core of family life was reshaped. Dual-income households appeared first to support a newer, better lifestyle but then steadily became a necessity for survival as families were caught between the changing structure of employment and rising expectations.

The evening meal with the entire family gathered together around the kitchen table has virtually disappeared from millions of American homes over the past twenty years. Restaurant expenditures now exceed one-half of all food purchases. Though that does not mean that half of all meals are eaten away from the home, it does reflect a major shift in eating habits. When combined with increasing expenditures for foods prepared outside of the home—home replacement meals—the claim that cooking is becoming a mere hobby in the typical American household takes on increasing credibility. Little of what is served on our dining-room tables entered the house as a staple ingredient.

It was once possible to extract a fair picture of American cuisine from the Department of Agriculture's figures on civilian consumption of basic foodstuffs (see Table 9.1). That is no longer the case. Knowing that the average American consumes 115 pounds of red meat per year tells little about the actual dishes that were consumed in this age of rapid cuisine change. The following discussion of contemporary American food consumption starts with an exploration of the general pattern of contemporary dining and then follows with an examination of changing patterns of the ingredients that compose our meals.

The Context

The very way that Americans dine has been revolutionized over the past thirty years. The never-changing menu schedule that shaped the tempo of traditional American meals has virtually disappeared. The six o'clock (or some other standardized time) dinner hour too disappeared as women moved

TABLE 9.1 Apparent Civilian Food Consumption: 1970, 1994

	1970	*1994*
Total meat	177.3	193.5
Red meat	131.7	114.8
Poultry	33.8	63.7
Fish	11.7	15.1
Eggs (#)	284.0	167.0
Dairy products	563.8	586.2
Fats and oils	52.6	66.9
Animal fat	14.1	11.6
Vegetable oil	38.5	55.2
Flour and cereal prod.	135.6	198.7
Fresh fruit	101.2	126.7
Canned fruit	23.3	18.3
Fresh vegetables	85.4	113.9
Canned vegetables (exc. tom.)	34.3	29.2
Tomatoes (canned)	62.1	75.3
Frozen vegetables	16.6	21.6
Caloric sweeteners	122.3	147.6
Coffee (gal.)	33.4	21.1
Soft drinks (gal.)	24.3	52.2
Beer (gal.)	18.5	22.5

NOTE: Pounds per capita except where noted.
SOURCE: U.S. Department of Agriculture.

into the workforce in unprecedented numbers. Almost 60 percent of all women of working age were employed in 1990, the highest percentages being in the prime child-rearing age groups. More than a quarter of all children under five spend their days in a commercial child-care facility, and almost half are cared for by someone other than the mother during the day. The impact of these changes on dietary rules and patterns has been enormous.

A respected food columnist in the 1950s commented that one of the great fears of children at that time was visiting friends at mealtimes and being forced to eat "funny food, like lasagna or tuna fish casserole, or some strange vegetable" (Collins, 1994). Inconceivable today, this statement captured the essence of the typical American attitude toward food. Children rarely ate food that was prepared by someone outside the family. Lunch was typically carried to school in a brown bag or metal lunchbox or children returned home to eat lunch with their mother and siblings. Families rarely ate in commercial establishments and even less often dined in the home of someone who was not related.

This food security has largely been replaced by an unstructured, fast-paced feeding schedule. The traditional "normal" meal with Mom, Dad, and the kids sitting together at the kitchen table at the prescribed time and leisurely consuming a home-prepared meal while discussing the day's events has disappeared from most homes. Indeed, in millions of homes, no more than a meal or two a week is eaten at the table with everyone present. Some have none at all. The concept of Dad always sitting at the head of the table and Mom at the foot is alien to many children, and the idea of using mealtime for relaxation and family bonding is almost inconceivable.

Breakfast has always been the most informal meal in the American household, and that trend has continued. Family members are operating on different schedules more than ever. Children often feed themselves as they get ready for school or day care. Ready-to-eat cereal has become the single most common breakfast food because even a five-year-old can prepare it without physical risk. Almost half of all Americans eat cereal preparations for breakfast, most of that cold cereal served with milk. Only about 10 percent of Americans consume a traditional breakfast of bacon, eggs, and toast, and most of those are over the age of forty. Almost 5 million children participate in school free-breakfast programs, which ultimately also has affected and will affect their decisions about what to eat for breakfast.

The invasion of the quick-service restaurant has brought new breakfast alternatives for adults. Many a mom and dad rush the kids to school or day care and then swing by a quick-service outlet for an Egg McMuffin or an order of pancakes or a bagel and a cup of coffee. In a small southern town, the stranger who wants to find the best breakfast biscuit would do well to skip the drive to the courthouse square, circling until she sees the largest clump of cars; rather, the best biscuit will be found on the outskirts of town at the end of the longest line at the drive-through window.

The quick-service restaurant today provides 62 percent of all breakfasts eaten in restaurants. The breakfast segment now represents from 15 to 20 percent of total sales of quick-service outlets. Traditional major chains and independent restaurants each provide about one half of all quick-service breakfasts; this ratio is beginning to change as increasing numbers of price-sensitive seniors discover that they can as easily relax with their breakfast group in a McDonald's or a Hardee's as in a traditional sit-down restaurant. The second largest provider is the midpriced restaurant with 35 percent of its sales at breakfast time; the upscale restaurant garners only about 3 percent of sales—much of that on elaborate weekend brunch buffets. The breakfast business is so important in some chains that it is becoming a driving force in strategic planning. The Shoney's chain, for example, realizes 58 percent of all sales from its breakfast trade, and the International House of Pancakes chain receives two-thirds of its revenues from breakfast sales.

The midday main meal has virtually disappeared from American life except at holiday times. Serving lunch exactly at noon is also beginning to be challenged. Retail workers rarely have an opportunity to eat during the lunch "hour" because business is booming; many office workers have the same problem. Recent surveys have found that the typical white-collar lunch "hour" is 35 to 45 minutes in length with the food often consumed at the desk rather than in the lunchroom. Even schools, once a bastion of normality, routinely schedule lunch shifts beginning around 10:30 A.M. and continuing until about 1:15 P.M. to reduce their need for larger lunchroom facilities.

The quick-service industry has played an important role in changing the nature of the American lunch by providing inexpensive food quickly. The hamburger, french fry, and soft-drink combo has long been the favorite selection of many. The deep-fried chicken sandwich has been increasing in popularity in recent years in spite of concerns about fat intake. Quick-service food has become so pervasive and Americans so brand conscious that brand-name-food outlets now appear in food courts and as miniunits in a range of corporate, institutional, and governmental settings as the only source of ready-to-eat food. Whether this changing attitude of the consumer has come about because of advertising or experience, the contemporary consumer is reluctant to eat hamburgers, hot dogs, and other fast foods that do not have brand names. Rather than test unknown waters, more and more opt for the slightly higher-priced brand-name products, which they believe will be of a higher quality. It takes a virtual miracle for the unknown single-unit, quick-service restaurant to survive.

The monotony of the grilled and deep-fried foods served by the quick-service outlets has also increased the market for "better" foods for lunch among the growing numbers of workers who have the mobility and resources to leave their workplace for lunch. Gourmet hamburgers were one of the first concepts to enter the scene, but today the main thoroughfares and food courts near office parks (somehow those restaurants actually within the parks seem to be less attractive, possibly because the workers want to get completely away from their work environments) are littered with near-quick-service restaurants pushing all kinds of specialties. Most workers are restricted in movement by time constraints, and even one's favorite food begins looking a bit stale when it is consumed day after day. Bundling of lower-volume quick-service restaurants has also grown in popularity as multiunit operators attempt to lure consumers by providing greater variety in the same amount of space. Ultimately, however, all of these food concepts are doomed to face the *monotonous* label because all food offered tastes basically the same due to the way that it is cooked. An explosion in the number of cold-sandwich shops in recent years has been one of the most visible results of this monotony problem.

Lunches of traditional food continue to decline in importance, especially among the younger consumers. Raised on fast foods, most of these consumers perceive that these traditional entrées are too heavy for midday consumption. Although the sandwich continues to be the single most popular lunch item, the ubiquitous hamburger is declining in relative popularity, especially among upscale consumers. Fine-dining restaurants now serve more pita sandwiches; caterers are more likely to serve chicken or turkey sandwiches. The veggie burger is the fastest growing menu item among these retailers.

School lunch programs feed about 10 percent of the entire American population (25 million in 1992). Though sandwiches are a part of the school lunch menu, pizza is routinely voted as the favorite meal. Tacos, chili, and an assortment of other foods have also become important parts of the school lunch as program administrators have worked to reduce waste and increase actual consumption of the food served. Ultimately the school lunch program has become one of the government's most effective nutritional educational programs, as it has introduced the nation's children to foods that they would never have been served at home. Though often criticized, the program has been a successful force in most districts in changing the nation's dietary habits.

It is impossible to characterize the American dinner because families scurry in seemingly endless directions. Preparation time has become a key element in the selection of entrées for the home-cooked meal. For example, the American Beef Board currently spends millions of dollars advertising how quickly beef dishes can be prepared and gives examples of exciting meals that can be cooked in under a half-hour. Similarly, one of the most popular cookbooks prints an estimate of total preparation time at the very top of the recipe beside the grams of fat in the meal. Dinner is the most likely meal to be cooked at home, though entrées and side dishes prepared in supermarkets, restaurants, and gourmet shops are one of the fastest-growing segments of the prepared-food industry. Home-cooked dinners made wholly from staple groceries are declining in importance at a rapid rate.

Holiday and special-occasion meals are the most elaborate meals served in most homes. The standards for holiday meals and entertaining have risen dramatically as home and cooking magazines have presented ever more exotic recipes and entertainment ideas, inevitably reshaping these events. The

Rural produce stands provide farm-fresh food to urbanites. (Richard Pillsbury)

concept that "presentation is everything" is reflected more and more on the festive home table.

The character of the typical evening meal continues to be largely determined by the economics, regional affinity, education, and cultural heritage of the family. At first glance the meal looks much like it did fifty years ago, but there have been significant changes beyond the increased role of meals from a box, the freezer, and the take-out window. Hamburgers, spaghetti (the sauce likely from a jar), and some form of simple chicken dish have become some of the most commonly recurring entrées. More than 600 million pizzas are delivered (or carried out) each year. There are far more vegetables served than ever before, though that fact may not pertain to meals served in the nation's most traditional households. All things being equal, the poor continue to have the least varied and nutritious diets even when better foods are available, because their meal regimes are the most traditional.

The economic and educational elite have the most-changed diets. Almost fixated with lowering fat intake, this group eats less and less beef, rarely consumes pork, takes it as a badge of honor to never use a frying pan (except for stir fry), and eats half again more fresh vegetables and twice as much fresh fruit than does the population as a whole. This group is also the primary consumer of fresh fruit juices for reasons of taste and health but simultaneously consumes far more soft drinks, alcoholic beverages, pastries, and candies than the population as a whole.

Most striking, the economic elite have the most "modern" meals of any segment of the national society. The primary purchasers of women's magazines and cookbooks, these are the Martha Stewart wanna-bes. They are the first to try innovative dishes for both everyday cooking and holiday occasions. Because they cook the least, cooking is more of a hobby. Rather than being bored with the monotonous routine of preparing most meals at home, they eat half or more of their meals in restaurants each week. The remaining home meals are an adventure, an opportunity to try out the newest recipes appearing in *Sunset Magazine* or *Bon Appétit*. They are also in the economic position to go out for dinner, and the garbage disposal consumes the home-prepared dinner when it is not up to their standards. Typically little ethnic cooking takes place among these consumers—hence the popularity of ethnic restaurants. The recipes that they do prepare, however, often blend traditional and new ingredients and methods of cooking.

The number of dinners eaten in commercial establishments increases each year. The traditional midpriced market grill or café, which represented the bulk of all outlets fifty years ago, has fallen on hard times as savvy consumers look for individual dining treats and theme meals. A typical large city has several thousand restaurants offering every food imaginable. Selecting among them is becoming increasingly difficult. Consumer restlessness reflects a growing ennui with the available choices. One recent consumer survey found that more than 70 percent of all suburban diners do not know where they are going to eat when they head the car out the driveway—the decision is made on the way to the restaurant and heavily influenced by traffic, lines in front of the establishment, and, not infrequently, intense last-minute interpersonal negotiation.

The attempt to make restaurant food palatable to all consumers has brought a general "evening" of flavors so that every entrée tastes pretty much like every other one. There may be a new restaurant every week down the street in suburbia, but most of these are pretty much alike. It appears that the public is beginning to rebel while simultaneously patronizing these establishments ever more frequently because of the predictability they provide.

It is interesting to observe multiunit operators flocking to seminars on location, location, location or learning how to squeeze the last penny out of back-store operations while never recognizing that if the customer is not made comfortable when he enters the store, he is never going to establish a bond with that place. There seems to be little recognition that the manager or maître d' used to actually perform an important service that cannot be duplicated by the vacant-eyed, properly pert, blond hostess who inevitably seems to work in the vast majority of midpriced and fine-dining restaurants today. For example, five mornings or more a week my wife and I drive fifteen minutes to a café with reasonably good food, but the decision to have breakfast there has more to do with the sense of comfort this place exudes. A good part of this sense is created by a decor that features familiar diner-style red booths, Formica-covered tables, and dozens of contemporary folk art pieces on the walls. But just as important in creating the mood for this restaurant, which may serve as many as 1,500 meals per day, is Kevin, the general manager, who recognizes us as regulars, greets us by name, and knows where we prefer to sit. The waitresses know our preferences and indulge us if we want to deviate from the printed menu. They also know that my wife will need at least eight to ten refills of coffee as the meal progresses. This store has learned that making the consumer feel comfortable sets it apart from the store across the shopping center parking lot, which serves much the same food to far fewer customers. Although ultimately we all go to a restaurant to eat, it is the ambiance of the place—the way the food is presented and the sense of well-being one has while dining—that transforms a basic body-food experience into something special and brings one back on another day.

The range of modern restaurants suggests that the American diet is changing very rapidly. The recent menu survey by *Restaurants and Institutions* magazine clearly indicates that modern menu additions have changed traditional eating habits. Cheesecake was found to be the top seller, followed by Caesar salad, french fries, and orange juice. In 1995 the top-selling entrées were grilled chicken breast, pasta, stir-fried chicken, and prime rib. Even more variations appear, however, when the menus are examined by type of restaurant.

Fine-dining restaurants have the most sophisticated menus, and many of their top sellers were virtually unknown in the American market a decade ago. Whereas filet mignon continues to be the top-selling beef cut, chicken and fish have virtually taken over the entrée list. Stir-fried chicken, grilled chicken breast, chicken or turkey parmigiana, empanadas, and salmon are the top entrée items. The top appetizer is calamari, the top pie is rhubarb, and the top salad dressing is house.

The family restaurant most likely uses pollack, without identifying the species, for its fish entrée and serves chop suey as the Asian specialty, spaghetti as the Italian, and fajitas as the Mexican; roast turkey is the poultry entrée. The salad dressing of preference is ranch, the pie is apple, and the cake is cheesecake. The similarly priced casual-dining restaurant, which typically is a chain venue, is a cross between the previous two. Grilled chicken breast, prime rib, fajitas, oriental (teriyaki) chicken, and crab cakes are the most important entrées.

These figures suggest that Americans are more adventurous in restaurants than at home. Part of this stems from the fact that it seems more appropriate to try something new in an unfamiliar environment; part stems from not having the ingredients or the time or not knowing how to prepare these strange dishes. Even institutional food purveyors have begun to join the exotic-food bandwagon. Cajun shrimp, dim sum, pot stickers, and bagels have all become regulars on hospital menus. The most conservative institutions continue to be schools and universities, but even these are beginning to change.

Ethnic cuisines have been at the center of the most exciting cuisine innovations in the United States in recent years. Most larger urban centers now have restaurants featuring thirty or more ethnic cuisines ranging from the obvious (Chinese, Mexican, and Italian) to the truly exotic (Guatemalan, Mongolian, and Nigerian). The most widespread emerging cuisines today include regional French and Italian, a selection of Mediterranean (including north African), and regional Chinese. The most common continue to be Italian, Mexican, and Chinese (Cantonese). The impact of this revolution is not restricted to dedicated ethnic restaurants, however, as their presence promotes the evolution of fusion cuisines—foods created from more than a single origin—and creates a sufficient market for more and more exotic ingredients in supermarkets for home consumption.

Missing Meals: On Obsolescence and Changing Preferences

The disappearance of favored foods and meals is also an important part of a people's food history. The addition of new dishes over the past 300 years has been chronicled, but little has been said about those foods that we don't eat anymore. Much of the colonial diet disappeared during the technological revolution of the nineteenth century, which brought us cookstoves, refrigeration, and prepared foods. The vegetable stew, the potage, and the pease porridge virtually disappeared from American tables as technology brought sufficient food reserves to allow excess food to be stored or thrown away. Salt pork, the heart of the colonial diet, has virtually disappeared from stores, as has salt fish. It is doubtful that even 1 percent of today's cooks have ever seen salt fish and salt pork, much less prepared them. Cornmeal, the primary cereal, too has fallen on hard times in mainstream kitchens, and cornbread has virtually disappeared from the nation's tables, though it appears in a much reduced role in the South. The disappearance of these foods has also meant that many of the dishes made from them are no longer included in home menus. It would probably be difficult to convince much of the population to even taste salt cod and fatback today, much less consume them on a regular basis.

Reflective of the rapidly changing diet is that many of the replacement dishes for those colonial foods have met the same fate as their predecessors. Prepared breakfast cereals, hamburgers, hot dogs, and the like have remained important, but many other replacement foods of that period have fallen into disfavor. Game declined in use throughout the nineteenth century and virtually disappeared from the dining table after World War II. Liver and onions were consumed regularly to increase the iron in one's blood, and fried chicken gizzards were a favorite when available; such organ meats are rarely consumed now. Boiled chicken is another victim of changing food preferences. Whole chickens are rarely offered in most supermarkets; in fact, most chicken is cut and packaged in processing plants that ship skin, fat, and other waste products to other processors for nonhuman food use. Meat loaf, part of the post–World War II weekly meal rotation, has virtually disappeared from the home because of its long cooking time. Declining pork consumption has brought a decline in sales of pork roasts and chops, also regulars on many home tables in the past. Homemade gravies are becoming rare as well.

One of my children not long ago referred to the meals of her contemporaries as box and bag meals. There is a large degree of truth in this. Canned foods are consumed less and less even though prepared foods are even more important. Most young families were raised in the quick-service era when there was no question that a young wife would work after marriage and interest in home cooking had typically declined among young marrieds. The traditional Sunday dinner, which took hours to prepare, has virtually disappeared. Dining on a standing rib roast or ham with mashed potatoes and canned green peas is something one does at Grandma's home on the rare visit, not at home on a regular basis. Indeed, many traditional vegetables are rarely eaten now; fresh green peas, carrots baked with the meat, cabbage, and turnips have largely been replaced by "modern" broccoli, zucchini, and lettuce salad. Macaroni and cheese, once a family entrée regular, now comes dried in a box and is used as a lunch or dinner side dish; tuna-noodle casserole has become passé, and even Ricearoni is increasingly looked upon as old-fashioned.

Restaurant favorites have possibly changed even more dramatically than home-cooked meals. Typical blue-plate specials prior to World War II included hot beef or pork sandwiches, meat loaf, chicken gizzards, and liver and onions; these meals are found today almost solely on menus in retro diners.

There was a time that *every* restaurant in the Northeast offered "spring" chicken and every restaurant of the Midwest had a pork tenderloin. These entrées have fallen victim to the changing dietary preferences and are perceived as old-fashioned. Even desserts have changed. Home-baked pies were once the single most common dessert in restaurants. A few chains continue to specialize in pies, but most do not even offer them. Those that are offered most likely come from Mrs. Smith or other factories scattered around the country.

The Content

Americans continue to be flesh eaters. The average American consumes about 115 pounds of red meat each year, though it should be noted that consumption is down from 134 pounds in the early 1970s and almost 200 pounds at the turn of the century. Beef is the most popular of the red meats at 64 pounds per capita consumption, though generally beef consumption has been declining since the early 1960s. Concerns over fat intake have been routinely expressed by American dietary gurus in recent years, and an increasing number of Americans perceive the consumption of red meat to be unhealthy. The nation has the largest number of vegetarians and partial vegetarians in its history.

Quick-service restaurants featuring the hamburger have played a crucial role in the continuing high volume of American beef consumption. It is impossible to estimate what percentage of beef is consumed in restaurants or how much is sold in ground form, but it is clear that consumption is dramatically increasing. Beef roasts and stew cuts are declining in sales as the American home cook turns away from recipes with long preparation times. The beef roast was a standard item in the mainstream family menu rotation thirty years ago; today's typical family consumes less than a beef roast a month. Stews are also not eaten as much, and when they do appear on the kitchen table, chicken is most likely to be the meat in the pot.

Pork consumption fell behind that of beef after the turn of the century but then exceeded that of beef during World War I and probably during the late 1930s. Beef and pork consumption after that were on about equal footing until after World War II. Pork consumption rapidly fell in the late 1950s but has remained stable for the past twenty-five years. Fears of heart disease and cholesterol are generally given credit for the decline in pork consumption, but this may be an important factor only among the nation's more educated population. Changing breakfast habits have played a far more important role in the decline of pork; the traditional breakfast of fried eggs and pork product with toast has given way to cold cereal, pastries, and other cereal breakfast products. The traditional American breakfast accounts for only about one in six breakfasts, and the percentage of those meals served without side meat (pork) is increasing.

Pork consumption has actually declined at an even greater rate than appears among the traditional population. Many of the 22 million people who have emigrated to the United States since 1965, especially those from Central America, Asia, and eastern Europe, come from cuisine traditions that feature pork. Although the adult immigrants typically have not changed their dietary attitudes about pork, their children ultimately enter the mainstream and consume less pork than their parents. The African American population continues to have the highest level of pork consumption, due primarily to its southern heritage, but levels among "other" racial categories are rising.

Poultry has been the growth segment of the flesh market. Concerns over total fat intake generally underlie this rise in poultry consumption, though this may be true only among the more educated sectors of the population. Typically the flesh market is polarized between the vegetarians and the poultry and fish consumers and the traditional beef and pork consumers. There has been little change among the traditional red-meat consumers, but as their average age increases, their relative importance in overall consumption decreases. Growing steak-house revenues indicate a countertrend in this area.

Turkey is the most interesting of the poultry products and represents a classic example of the widely held belief that California sets the nation's cultural norms. Nontraditional turkey sales began increasing in California during the 1950s. Californians began purchasing turkey breasts in increasing numbers during this period to avoid having to roast an entire bird. Ground-turkey patties began appearing in California grocery cases in the early 1960s. Today ground turkey occupies nearly the same space in the meat cases of some markets as ground beef. Turkey sales have been slow to grow in the eastern United States, but turkey is becoming more visible in those western urban markets where the dominant grocery chain has significant sales. Although turkey sales are still small, only 14.4 pounds per capita, they have increased nearly 125 percent in since 1965.

Food nutritionists have made a strong case for seafood as a healthy source of protein, but their advice has had little impact on the American diet. Sales are still at only a few ounces per week per capita, most of that deep fried. Americans for the most part are reluctant to consume seafood regularly. They primarily consume fish with the least flavor when they do. Generally the consumption of fresh and frozen fish is increasing relative to all fish consumption; sales of canned salmon, salt fish, and shellfish are declining. Canned tuna continues to be the single most consumed fish product with a total weight of 3.3 pounds per capita annually. The fresh- and frozen-fish category is almost twice that amount (6.4 pounds) but represents almost two-dozen species.

Alaskan pollack is the nation's most consumed fish in fresh and frozen form, though few Americans are aware of ever eating this species. Marketers have long believed, despite no evidence that consumers have a negative image of pollack, that most Americans would shy away from consuming this fish. Most unidentified fish filets and fish sandwiches served in America are pollack. The 1991 wild harvest exceeded 2.75 billion pounds. In recent years the production of farm-raised catfish has increased sufficiently that protein companies are also using farm-raised fish to provide generic fish products.

Salmon is the second most important fresh- and frozen-food fish, though wild harvests vary widely from year to year as the National Marine Fisheries Service and other government agencies attempt to regulate future supplies. Farm-raised Atlantic salmon have been in the American market for more than a decade; they are primarily from Norway and Chile. The consistency of supply and quality of this farm-raised product has encouraged many independent and multiunit restaurant operators to feature it and the perennial favorite, fried shrimp, as their everyday fish entrées. Cod and flounder are the third and fourth most important wild-harvest fish. Crabs are the most important of the shellfish with an annual catch of 650 million pounds, followed by shrimp at about half that amount. Processed pollack often is used as a crab substitute for restaurant buffets.

Farm-raised fish are becoming increasingly important in the American diet because fish are even more efficient at converting grain feeds to protein than chickens. Salmon, oysters, and catfish are the three most well known farm fish currently being consumed, but there has also been a rapid expansion of harvests of farm-raised crawfish, tilapia, shrimp, clams, and even a few abalone in the past few years. Markets for several of these species, most notably catfish and tilapia, have been created where none existed previously through strong marketing campaigns. Whereas tilapia have been well accepted, the less attractive catfish has had a difficult time being accepted as a white fish outside the South. More than a half-million pounds are consumed primarily in fish sandwiches and as unidentified white fish filets. Catfish restaurants and fish camps continue to be popular in the central South, but catfish rarely appears on menus elsewhere.

The issue of fat content is largely a red herring in trying to understand the rapidly changing contemporary American diet. Annual red meat consump-

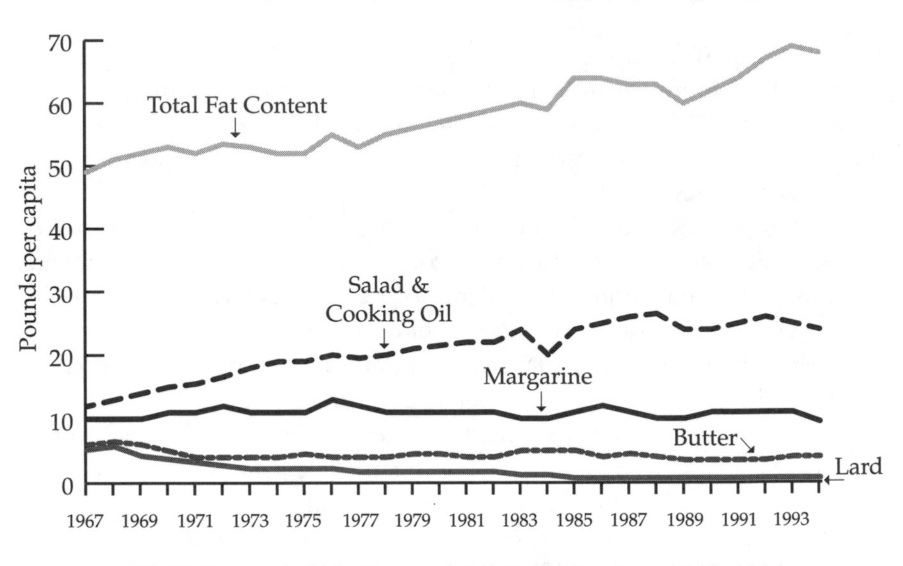

FIGURE 9.1 Edible Fats and Oil Consumption: 1967–1994

tion indeed has declined by seven pounds per capita over the past thirty years, but the consumption of cooking oils and other fats has risen by 21.8 pounds per person (see Figure 9.1). Admittedly there has been a small decline in animal-fat consumption, but total fat and oil consumption has increased. A part of this change is due to the increasingly segmented market; those attuned to a low-fat diet consume virtually no animal fat, but an even greater number of consumers continue to ignore the fat issue in their menu decisions. It would be easy to blame the french fry for the rising consumption of vegetable fats and oils, but it constitutes only a part of the problem. The increasing importance of quick-service restaurant dining, where large percentages of all foods hit the deep-fat fryers, also plays a major role. Ultimately the consumer continues to receive an inner satisfaction from the taste and texture of fat-laden foods.

Consumption of dairy products has undergone the greatest change during these past thirty years of any basic food on grocers' shelves. Total milk consumption has dropped by more than 25 percent to only 213 pounds per person annually (see Figure 9.2). The decline of so-called whole-milk consumption has been devastating for producers; sales have slid from 169 pounds to only 76 pounds per person per year. Part of this decline has been offset by rises in low-fat and 1 percent milk products. The consumption of skim milk, which seems to be an acquired taste for most people, has increased only slightly. This slide in milk consumption almost certainly will continue as the

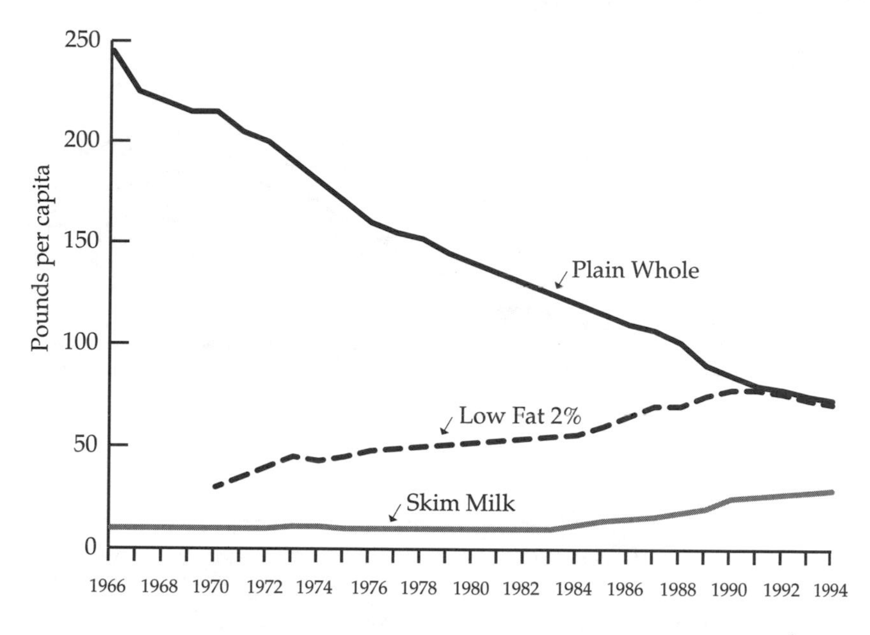

FIGURE 9.2 Milk Consumption: 1966–1994

milk-drinking European-origin population declines in relative importance, has fewer children, and generally is more concerned about fat intake.

Other dairy products present a much more complex picture; total dairy consumption is up significantly over the past twenty-five years. Most of that increase is from increased use of cheese as either a garnish or an additive (see Figure 9.3). Salad bars routinely offer at least one type of cheese as a topping; cheese soups and other dishes with cheese sauces are also common in many restaurant recipes. Sour cream is a very popular condiment for many Mexican foods, and it is not surprising that its use has more than doubled with the growing numbers of Mexican Americans and the current ubiquitous presence of quick-service, if not full-service, Mexican and Mex-Tex restaurants.

The pizza revolution is particularly reflected in the increased mozzarella cheese consumption, which rose from a little more than one pound per person per year at the beginning of the pizza era to eight pounds per person today. Cream-cheese consumption has similarly tripled with the increased popularity of bagels for breakfast and cheesecake, now the single most popular cake dessert in fine-dining restaurants. Consumption of American and processed cheeses has also increased a great deal during the past twenty-five years, partially because of the rise in Mexican food consumption.

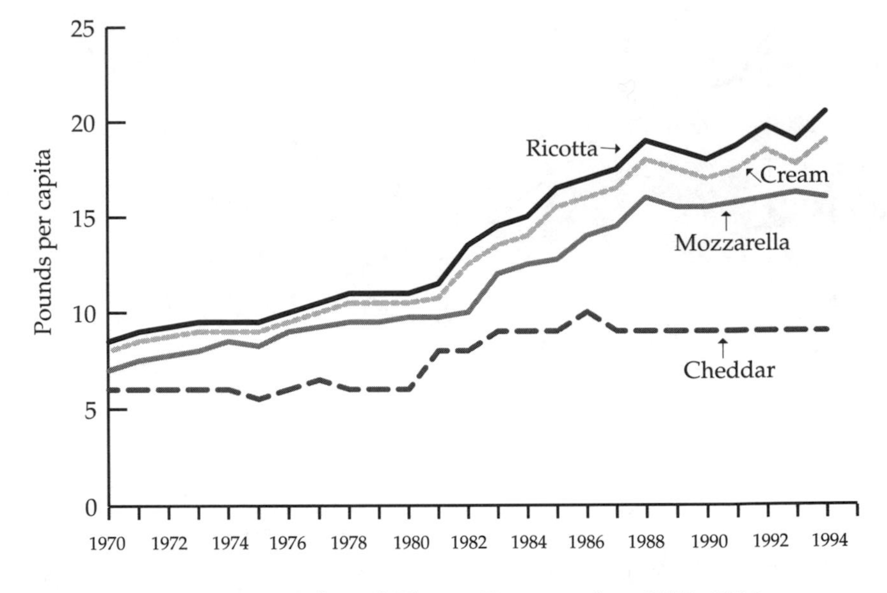

FIGURE 9.3 Selected Cheese Consumption: 1970–1994

Changes in the consumption of carbohydrates have been as dramatic as in the consumption of meat products (see Figure 9.4). Consumption of flour and grain products has increased in all areas over the past twenty-five years and is up overall by almost 50 percent. Bread consumption generally has increased because of the continuing importance of the perennial sandwich restaurant. Rice consumption almost tripled with the immigration of 5.5 million Asians during the same period. Cornmeal consumption also doubled with the arrival of millions of Latin Americans, though consumption in the general population has declined significantly. The 150 percent increase in durum wheat consumption is almost entirely related to the increased consumption of pastas and bagels.

The lowly potato has virtually been reinvented over the past twenty-five years. This mainstay of the American diet was primarily served boiled and covered with gravy through the 1950s. Changing dining patterns soon brought changes. The supposed favorite meal of every man—pot roast and mashed potatoes with gravy—evolved into a steak and baked potato stuffed with sour cream in the 1950s, partially because of the shorter preparation times of the steak, partially because of the rise of the inexpensive steak house featuring baked potatoes as its primary side dish and partially because of rising disposable incomes. Eating a steak and a baked potato at least once a week at home or out became a part of the national self-image in the 1950s; it

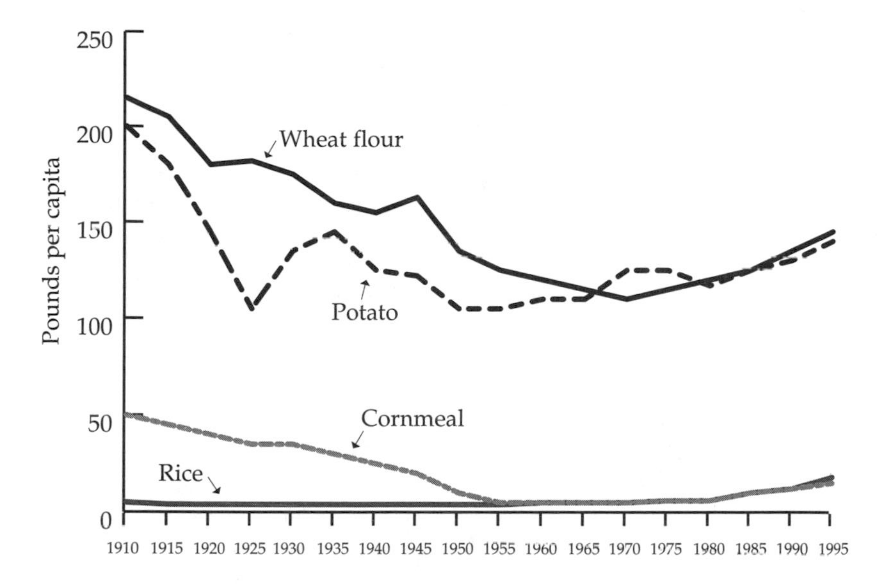

FIGURE 9.4 Selected Carbohydrate Consumption: 1910–1995

was one way in which the upwardly mobile middle class could identify itself as a part of the new American success story. Baked potatoes had been served only sporadically at home prior to that time and almost never in restaurants. The development of specialized restaurant equipment to prepare potatoes and keep them hot and the sense that potatoes and gravy were somehow a part of that inner-city past that the new suburbanites were trying to leave set the scene for a dramatic rise in baked-potato consumption both in restaurants and at home.

Simultaneously, quick-service restaurant operators saw the advantage of pushing the high–profit margin french fry to offset the lower margins of the inexpensive hamburgers that they were forced to offer in order to compete in a crowded marketplace. The new french fries were so good that not only McDonald's but all of its competitors quickly adopted them as their only fried potatoes. The new product could be thrown into the deep-fat fryer while still frozen and heated and browned in just a few minutes. McDonald's ultimately based its reputation for quality not only on product consistency and clean bathrooms but on serving the finest french fries in the quick-service business. McDonald's used lard for many years, the real secret behind why its fries tasted better than the competition's, but ultimately was forced to switch to vegetable oil because of public concerns over animal-fat consumption. The new french fries tend to be limp.

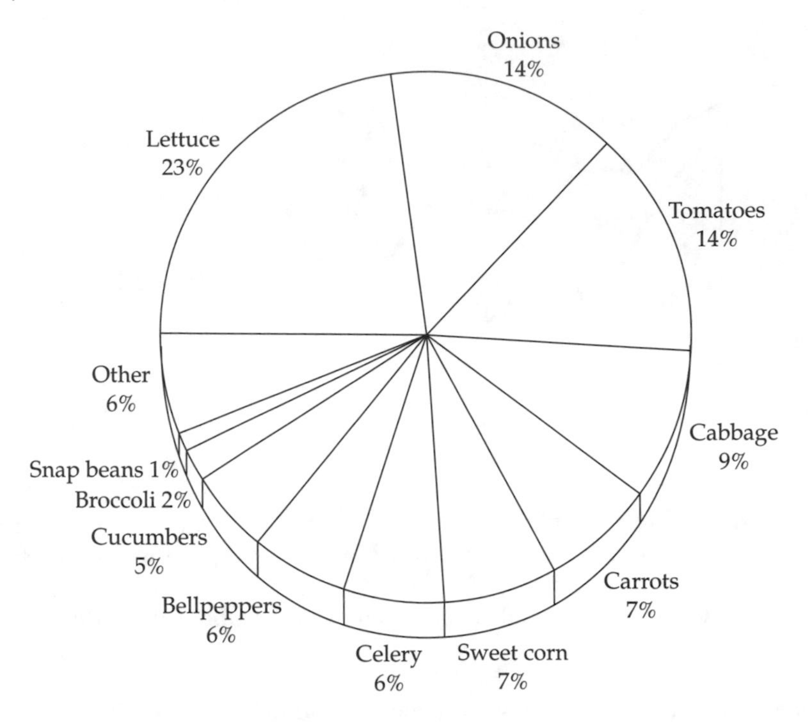

FIGURE 9.5 Fresh-Vegetable Consumption: 1994

The boiled potato almost disappeared from the American table after 1970. Although seemingly an innocuous change, this created a chain reaction in the geography of potato production. The potato varieties that have the soft, smooth consistency to make the best mashed potatoes grow well in the humid East; the russet variety, which has the proper granular texture for both french fries and baked potatoes, grows with the best texture in an arid climate under irrigation, as is found in the intermontane West. Potato production in the eastern United States declined sharply. In a matter of years Idaho moved from being a moderately important producer to becoming the "Potato State." It didn't hurt, of course, that J. R. Simplot's company was headquartered in Boise, Idaho.

The average American consumes a half-pound more vegetables each week today than in 1974 (see Figure 9.5). Fresh produce has been responsible for the bulk of the increase, the greatest growth coming from increased consumption of onions, green peppers, tomatoes, and carrots. The consumption of head lettuce, which accounts for one-fifth of all fresh-vegetable consumption, underwent little change, as the salad and salad bar revolution had become well established prior to 1970. What changed were the vegetables that were added to that lettuce to make increasingly complex mixtures. Fresh broccoli, cucumber, and cauliflower consumption all increased significantly; earlier traditional favorites, most notably radishes and celery, were used less often.

Frozen-vegetable consumption has undergone an almost 25 percent increase over the past thirty years. Sweet corn (which is now almost as good frozen as fresh) was the big winner, though broccoli and snap beans are also attracting consumers. Canned vegetables, in contrast, lost sales, especially if canned tomatoes (the Italian pasta revolution pushed their consumption dramatically upward) are subtracted from the consumption numbers. Canned sweet corn and peas were the most obvious losers, but virtually all canned vegetables fell into disfavor as processors improved their frozen product and shippers developed more cost-efficient methods of shipping fresh produce to consumers.

Fruit sales also benefited from the increased national emphasis on better nutrition. Average consumption increased by almost a full pound per week, about evenly distributed between fresh and processed products (see Figure 9.6). Banana consumption rose to an annual per capita consumption of twenty-six pounds, followed by melons at twenty-four pounds. Grapes increased dramatically in popularity after the introduction of the popular flame seedless variety, which shipped and kept better than earlier grapes. The concentration of almost one-quarter of the nation's total strawberry production in a single county brought massive economies of scale, which kept prices relatively low. The big loser was the orange.

The American thirst also underwent significant alterations as the nation's sweet tooth pushed beverage consumption away from old favorites and toward sweetened drinks of all kinds. The owners of Coca-Cola stock saw it soar 1,500 percent over the past twenty years as carbonated soft-drink consumption swelled from an average of twenty gallons per year in 1966 to a gallon per week in 1994. Cola increasingly became the flavor of choice, although regional preference patterns continue. Whereas it verged on being a social sin to drink carbonated beverages before noon in 1966, today it appears that almost as many soft drinks pass the counter of a quick-service restaurant during the breakfast period as the more traditional coffee in many parts of the nation. An interesting contradiction to this trend has been the virtual disappearance of the practice across the South of pouring a bag of peanuts into a bottle of Dr. Pepper and drinking the mixture for a midmorning boost.

Coffee has been one of the most obvious victims of changing beverage-preference patterns (see Figure 9.7). Coffee consumption declined by almost 15 pounds per person between 1910 and 1994 with consumption focused

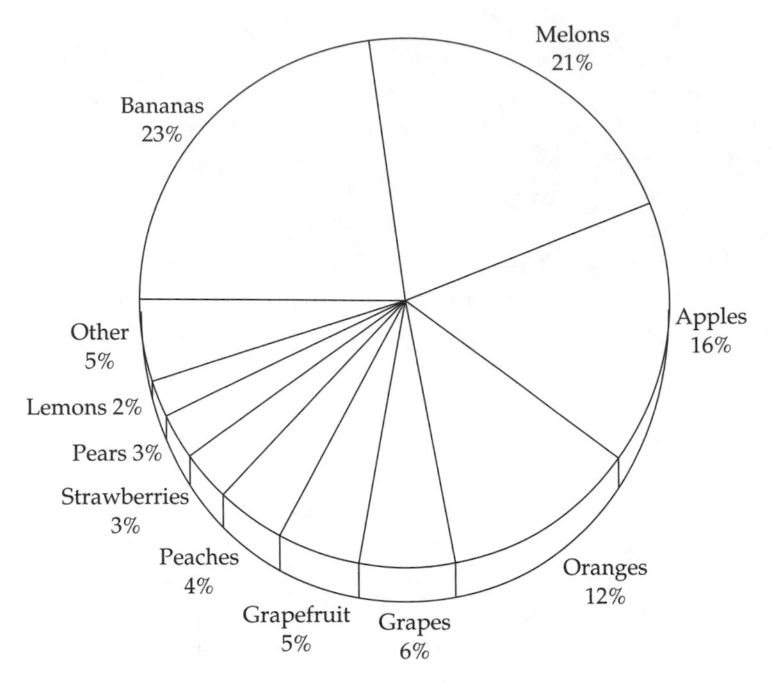

FIGURE 9.6 Fresh-Fruit Consumption: 1994

primarily among those over thirty. An interesting countertrend has been the rise of specialty coffees and the emergence of the new American coffeehouse. Despite increasing alcohol consumption generally, there appears to be a significant minority of younger people in need of a "third place," an away-from-home gathering place for meeting friends where alcohol is not the center of attention. The Pacific Northwest has long been home to the nation's heaviest coffee drinkers and the most coffeehouses. The Starbucks chain of coffeehouses, based in Seattle, has rapidly spread across the nation. The coffeehouse still accounts for only a few thousand units nationally, as opposed to more than 35,000 (alcohol) drinking establishments and a half-million commercial food establishments.

The nation's taste in alcoholic beverages also has changed dramatically (see Figure 9.8). Distilled-beverage consumption dropped by almost a third between 1950 and 1994 with the harsher-flavored whiskeys suffering even higher losses. Rum and vodka consumption increased, presumably because of their lighter flavors; consumption of cocktails and straight distilled beverages generally is seen less and less in social situations. Wine consumption, which was virtually nonexistent in 1950, reached 1.7 gallons per person over

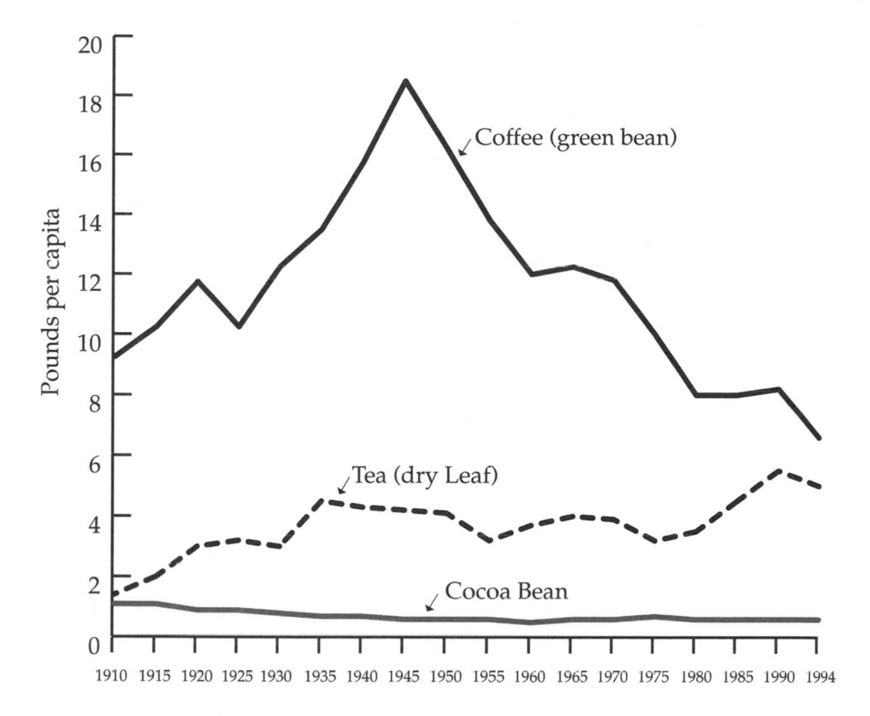

FIGURE 9.7 Coffee, Tea, and Cocoa Consumption: 1910–1994

twenty-one by 1966 and today amounts to almost two and a half gallons per year. The kinds of wine preferred have changed even more—from the light white wines (Chablis and Rhine wines) to the heavier reds (Bordeaux and cabernets) back to the lighter blushes (white zinfandel) and most recently toward increasingly exotic whites.

Beer drinking has taken up where distilled beverages left off. Not only has annual beer consumption increased by more than six gallons per person over twenty-one since 1950 but the kinds of beers consumed has also changed. The heavier batch-brewed beverage was replaced initially with a lighter, continuously brewed product in the 1960s. Lighter and lighter brews were introduced throughout the 1980s and early 1990s, ultimately opening a market for beers with more distinctive flavors as a countertrend. Anchor Steam beer from the San Francisco Bay area was the first of the new microbrews to obtain a national image, but it was quickly joined by a host of others. Brewpubs are the current fad in the industry, but like so many other fads in the food-service industry, it appears that their numbers will soon exceed their potential market.

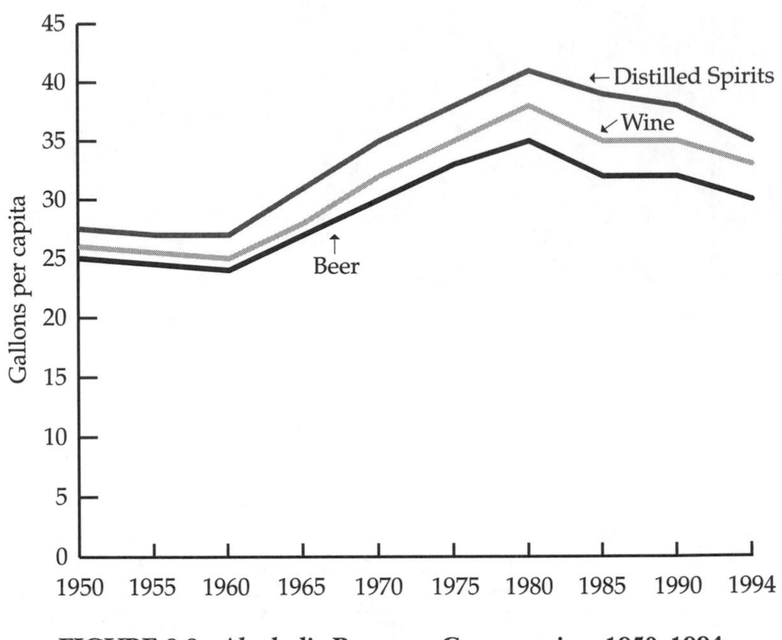

FIGURE 9.8 Alcoholic Beverage Consumption: 1950–1994

A Few Final Thoughts

There has never been a national menu or cuisine in America; rather there has been a broad set of food preferences largely based on the nation's western European roots. The details of this cuisine have changed dramatically since the beginning of the Industrial Revolution; yet the evolution of the new American menu has been so gradual that few have been aware of the changes as they have taken place. The new menu seems to us to be much like that of our grandparents and great-grandparents, but it is much different. Ultimately food preferences are about comfort, and the gradual amalgamation of foods from around the world has allowed them to be Americanized. Stir-fry and enchiladas, hoppin' John and black beans, have all been added to the nation's foodways over the past two and a half centuries; yet there still is no sense of alienation. Ultimately the vitality and power of the nation's all-embracing culture has meant that there can be no foreign food.

The drive toward a national culture has not destroyed the regionality of American food preferences, though it has changed the regions and how they interact. The next chapter focuses on the continuing regionality of American food, why it exists, and where it is going.

10

Cuisine Regions: Concept and Content

Things are more like they used to be now than what they were before.

—attributed to Mamie Eisenhower

The geography of American foodways has been almost totally transformed over the past century. Whereas most of us recognize that the content of our diets has radically changed, the geography of food preferences too has been transformed by the same forces; only shadows remain of the traditional regions. Just as we are proud of our historic districts, such as in New Orleans and Charleston, and of our re-creations of past events, for example, the reenactment of the minutemen on the green in Castine, Maine, there are parts of America that take pride in their continuation of a traditional cuisine. Closer inspection, however, reveals that these foodways are only as real as Williamsburg Village. Yes, they seem authentic, but they have been re-created and, in the process, enhanced. Like Williamsburg, they are not a part of the natural progression of heritage or the continuance of a way of life but are composed of some real elements and some created ones that represent our notion of the past.

Almost 10 percent of the American population has lived in this country for thirty years or less. The question of English as the only national language has reared its head for the first time in almost 100 years. Indeed, only the South has not been invaded by millions of immigrants since 1820, and even there change is beginning to take place as the spreading Caribbean and Hispanic wave of immigration is beginning to be felt, to say nothing of the even larger numbers of retirees and corporate gypsies who have moved to the region from elsewhere in the United States to enjoy its salubrious climate.

The nationalization of the American cuisine seems at first glance to have been universal. McDonald's and Pizza Hut seem to have invaded every community no matter how isolated. And indeed, it is possible to obtain most contemporary foods in every community regardless of size. This does not mean that the nation is suddenly homogeneous and that regional patterns do not exist; rather it suggests that each year the traditional repositories of our regional images are less and less like those that reside in our collective memory. It was once easy to determine when one crossed a boundary from one place to another when traveling by highway or even by air. Though there have always been transition zones, once in the new area one could easily see that one was now in a new place. That is becoming increasingly difficult today. The invasion of standardized signage, corporate retailers, and international manufacturers as well as a highly mobile population and the general placelessness of most of urban society has meant that the connection with the past is just not as strong as it once was.

Visual keys to place are especially difficult to discern in this milieu of national retailing and structural design. Yet we know that regional variations still exist. In a recent trip across central Texas, we found a culinary landscape dotted with corporate franchises and supermarkets filled with the products of international food processors. We also ate a meal of beef fajitas and homemade guacamole prepared under a great oak in the parking lot of a new supermarket opening for business in Rocksprings. What made the fajitas special was not the ingredients, which came directly from the shelves of the new store, but the added flavor of heritage: There is a long tradition there that making the stranger feel welcome is as important as feeding his belly.

The goal of the following discussion is to illustrate the continuing presence of regional differences—differences, however, that are only casually related to the earlier colonial patterns. In the final analysis we can never truly understand the totality of our foodways, just as we can never totally be in tune with ourselves, but we can learn something of their temporal and spatial character. In the following discussion I first propose a model for cuisine regions and then attempt to determine how well it fits the complex reality in which we live.

Concept: A Regional Model

No two Americans have exactly the same taste preferences or eat the same foods. Similarly, any attempt to make regional generalizations must ignore many individual and local peculiarities that we know actually exist. The focus of American life has shifted dramatically away from the traditional eastern power and cultural centers over the past fifty years to create a new distribution of preferences, life patterns, and geographical associations. This is no more true than in the nation's foodways.

Although it is possible that we tend to oversimplify the geography of past foodways, it does appear that contemporary foodways are far more complex. The contemporary dietary regions do not lie over America like so many solid blankets, as was true of their traditional predecessors; rather each of the regions is more like a patchwork quilt created from distinct elements: the traditional fabrics, the historic transaction zones, and the national interaction zones. Today the nation is undergoing a rapid digestion process that simultaneously involves the continued social restructuring of the native-born popu-

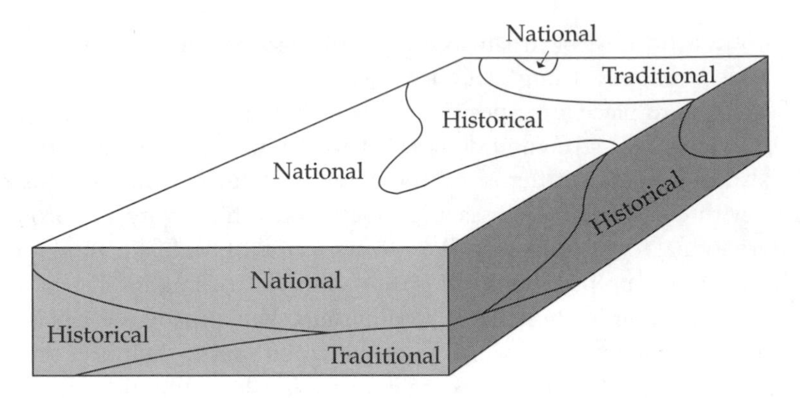

FIGURE 10.1 Components of the Regional American Diet: A Model

lation, including the incorporation of the new Americans, and the general conservatism about changing foodways. The result is a complex cuisine milieu in which virtually every single place simultaneously incorporates elements of both the past and the present to create strong images and complex foodways (see Figure 10.1).

The Traditional Fabric

The traditional fabric is the foundation of all contemporary regions. It is not untouched by modernity, of course, for nothing is static. There are few kitchens in America without a can of Campbell's soup, a package of frozen vegetables, or a box of Bisquick. Change is present, but the changes have taken place at a slower pace than elsewhere. The most obvious changes are more varied diets and a general decline in the importance of seasonal menus. Some ethnic foods not available during the evolution of the traditional diet are now present, but mostly they appear in the Italian, Chinese, or Mexican restaurant or at home when the occasional meal of a little spaghetti with Ragu sauce is served.

The traditional fabric once covered the entirety of the eastern seaboard and much of the Rio Grande valley, but today it is little more than a patchwork quilt. In some areas the traditional fabric seems to dominate contemporary life with only islands of change taking place, as is seen in much of the South and Appalachia. In other areas the traditional fabric only peeks out around the corners, as is true in New England and New Jersey. Nowhere is it quite like it once was.

Historic Transaction Zones

The historic transaction zones are portions of the country where sufficient in-migration occurred during the nineteenth century that the integrity of the traditional fabric was jeopardized. Typically these places accepted large numbers of immigrants from eastern and southern Europe to their labor-starved factories during the height of the Industrial Revolution. Other examples are central Texas, where large numbers of German and central European farmers settled around 1850, and the northern Midwest, where tens of thousands of Scandinavian farmers settled after 1870.

These places have much in common with the traditional subareas in that their foodways fall somewhere between those of the past and contemporary national standards. Like the traditional areas, they have been evolving toward homogenization, just more slowly than more mainstream areas. Homogenization has also taken place within these places as the differences between the foodways of the various ethnic groups have inexorably become blurred, especially in smaller communities. It has become increasingly difficult over the years in the mining and factory communities of the industrial Northeast to determine the ethnic origins of families by the foods that appear on their kitchen tables. Intermixing and intermarriage have melded the daily life of many into a sort of standardized industrial-era immigrant lifestyle. Spaghetti, kolbasa, goulash, stuffed cabbage, and other ethnic items seem to be consumed by everyone in the coal camps of western Pennsylvania regardless of their ethnic heritage. Potato pancakes, pork, and rich Scandinavian and German desserts seem to be endemic to traditions of the historic zones in the northwestern Midwest regardless of the ethnic heritage of those families. This does not suggest that these people are not proud of their specific heritage, rather that some of the ways of others have entered their daily lives.

The best examples of the historic interaction zones have been largely passed over by recent economic expansion. Out-migration, rather than in-migration, has been the order of the day. Few strangers settle in these areas to introduce new foods, few chain restaurants are drawn to their stagnant economies, food stores are aging, few new superstores with large stocks are replacing them, and purchasers with an eye to national trends do not live here. The foods here are as static as those of the traditional fabric zones; they are just a bit more advanced in time and a bit more focused in ethnic origin.

National Interaction Zones

Some portions of the nation have been the focus of enormous growth since World War II. Primarily occurring in cities such as Los Angeles, Dallas, Denver, and Minneapolis, this growth is not entirely confined to urban areas. The Sea Island coast of South Carolina, much of southern Florida, and the Palm Desert of southern California also have undergone massive growth and are classic examples of nonurban areas where the new order reigns supreme. Urban or rural, these areas have embraced the new American cuisine in such a way that the traditional foodways that have survived continue more as trophies than as real preferences. For example, Frogmore stew and clambakes are held up as emblems of the past rather than as markers of the present.

The classic cuisine of the national interaction zones has little or no regional, ethnic, or historical identification with the areas that surround them. The food preferences of those who live there are focused on the latest nutritional and flavor fads, and devotees engage in a restless search for new taste treats, new restaurants to sample, new concepts to test. The residents of these areas have the highest rates of dining away from home and the lowest rates of consumption of basic staples. Most know little about their past and are happy with that condition.

Content: Contemporary Diet Regions

The new regions reflect the changing economic order with only hints of past patterns. The northeastern seaboard was engulfed by the immigrations of the Industrial Revolution; eventually the colonial coastal cities amalgamated into a single economic zone that stretched westward until it met the coal fields of western Pennsylvania and the industrial Midwest. The Midwest evolved during the technological and agrarian revolutions into a dichotomous lifestyle of rural and urban that has begun to blur in recent years as rural life has simultaneously collapsed and expanded, forcing thousands into the cities. The South is the least changed of the traditional areas, although recent amenity-driven migration around the nation has brought the most attractive communities and rural areas into the modern nationalizing era. The West was once a place devoted almost entirely to ranching and farming, but in recent years the expanding Hispanic migrations northward and the economic explosion along the Pacific Coast have created three distinct western

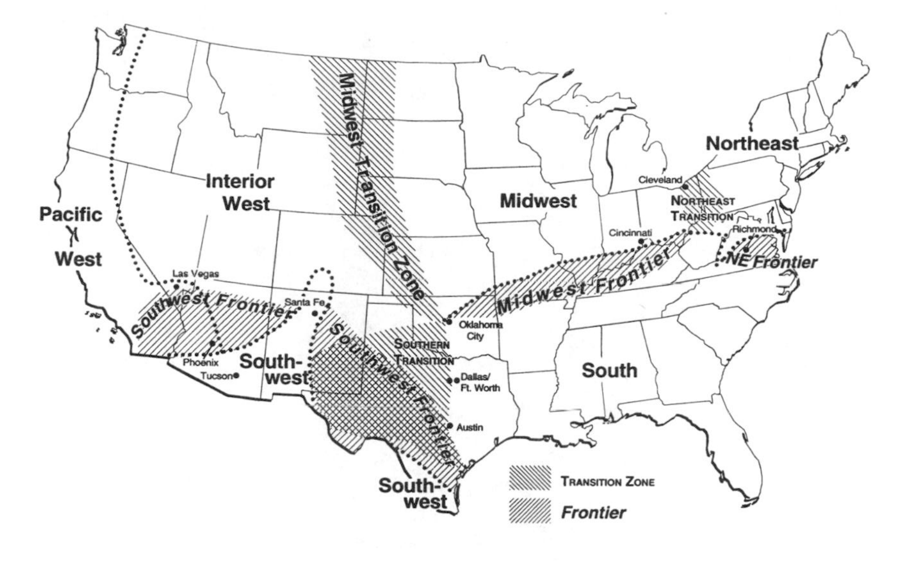

Contemporary Diet Regions

regions. But all of the patterns are far more complex than this and a more complete discussion of each of the major regions is necessary to understand their distinctive and similar qualities.

The Northeast

The Northeast easily contains the most varied sets of foodways in the United States today because of its rich colonial tradition, complex immigration history, and highly localized zones of growth over the past forty years. A megalopolis, the urbanized zone stretching from southern Maine into northern Virginia, is the heart of this rich demographic soup. Millions of second- and third-generation descendants of eastern and southern European immigrants are scattered throughout the entire region and a part of almost every industrial community. Millions more southern rural migrants began streaming northward in the 1930s and became an avalanche of humanity after World War II. Concentrating in the urban cores of the largest cities, these mostly African American migrants have begun to spread into the smaller cities over the past decade or two, following the factory jobs that brought them in the first place. Hundreds of thousands of Caribbean-origin restaurants have been added in recent years. Originally of Puerto Rican origin, these central-city residents are as likely to have originated in Jamaica or Trinidad as Puerto Rico. Last, there are the newest immigrants still piling off the boats and air-

Contemporary Diet Regions: The Northeast

planes. Most likely to be from Southeast Asia, eastern Europe and the former Soviet Union, and increasingly even Africa, these newest residents are almost entirely concentrated in the cores of the largest cities.

The descendants of the eastern and southern European immigrants are the most numerous throughout much of the region, and the basic modern diet is heavily influenced by their preferences and tastes. The favorites of these descendants are mainly those of their ancestors—Italian pastas, stuffed cabbage, rich meat dishes—and there is a general avoidance of most green and yellow vegetables. These diets have changed the least, especially in the middle and lower economic groups, and represent the most conservative foundation of the region's cuisine. Most of the other groups continue to be largely concentrated in ethnic residential zones, where their diets are going through the Americanization process.

Rapid nationalization of the cuisine of the eastern megalopolis is also taking place on the suburban peripheries among the children of all of these residents. Not as nationalized as suburbs elsewhere in the nation, suburbia here is undergoing increasing pressure from national chain restaurants and food stores. McDonald's and the other quick-service providers were slow to enter, but once they gained a foothold, they became as prolific in the urban cores and peripheries as anywhere in the United States. In general the midpriced

TABLE 10.1 Consumption of Selected Foods by Region: 1987–1988

Item	*All*	*NE*	*MW*	*South*	*West*
Beef	77	67	79	86	69
Pork	42	41	40	49	32
Lunch meat	20	16	23	22	14
Poultry	64	71	55	67	64
Fish, shellfish	22	27	17	25	20
Cheese	99	101	105	88	105
Shortening	3	2	3	5	2
Salad, cooking oils	5	6	4	7	4
Flour, not in mixes	11	7	10	14	13
Sugars	23	17	19	34	16
Ketchup, chili sauce, etc.	7	9	7	6	6
Vegetables (fresh)	120	128	102	122	134
Vegetables (canned)	39	38	37	47	29
Vegetables (frozen)	12	15	10	14	9
White and sweet potatoes	65	61	70	70	56
Fruit (fresh)	147	144	143	137	172
Fruit (canned)	10	10	13	9	9
Frozen fruit juice	49	41	60	38	64
Fresh fruit juice	36	64	25	37	18
Fresh fluid milk	297	304	337	250	315
Tea	2	4	1	3	1
Soft drinks	174	164	194	179	150
Alcoholic beverages	57	56	52	53	72

NOTE: Indexed based on 21 meal equivalents per capita. Based on survey data for 1987–1988.

SOURCE: U.S. Department of Agriculture.

sit-down national chain restaurant is underrepresented in suburbia, but the slack is more than taken up by traditional independents who continue with local favorites while offering more and more national dishes. A suburban diner today is more likely to offer grilled chicken breast than stuffed cabbage as the daily special but typically will offer both to keep its clientele satisfied. Reflective of this is the almost doubled consumption of frozen entrées in the region, along with the gradual replacement of beef by chicken and fish and the 50 percent increase in carbonated-beverage consumption in the home over the past decade (see Table 10.1).

The Washington metropolitan region lying at the end of the megalopolis corridor is the most truly nationalized subregion in the Northeast. Almost entirely a product of the postwar culture, suburban Washington, D.C., has poured out of the District of Columbia southward beyond Fredericksburg, Virginia, and westward into the Shenandoah valley of West Virginia. There are attempts to incorporate the past of the region into its contemporary foodways, especially in the ever present visitor service industry, but in actuality there are few here with historical ties to this place. As in much of the West, almost everyone came from somewhere else and is interested only casually in what was there prior to their arrival.

Historic New England is largely gone today, replaced by a mixed recreational-pseudoagrarian landscape that has been largely reshaped by national trends. Economically and culturally dominated by the larger cities to the south, this nationalizing zone is quite different from similar areas in that an important element of the modernization process has been to place a historic veneer over new and old alike. The cuisine is a combination of contemporary national fare and "nouveau" traditional fare. The image of historic New England is so important to the economic foundation of the region that it has become a regional signature to include a variety of updated historic recipes in almost any menu that is presented to outsiders. Inevitably these new approaches to old favorites enter the home-cooking milieu as well.

Southeastern Pennsylvania and some of the less settled parts of New Jersey are caught in much the same mind-set. A significant part of the local economy is based on fulfilling the ever increasing demands of urban foragers who search this countryside for recreation and a sense of historic place and buy farm-fresh produce for their tables. Providing "Pennsylvania Dutch" experiences to tourists, and increasingly residents, has become a major industry in southeastern Pennsylvania; New Hope and similar communities in New Jersey provide much the same pseudohistoric connotations. It is difficult to determine how many of the traditional foodways continue in the homes of the residual traditional population, but inevitably traditional lifestyles are being buried under the schlock of the tourist industry and suburbanization.

The remainder of the Northeast continues to slowly evolve from its traditional dietary fabric toward a contemporary cuisine that is largely a product of the past. Most of northern Maine, upstate New York, central Pennsylvania, and northern West Virginia have been areas of out-migration for more than seventy-five years. The paucity of newcomers has kept the introduction of new approaches to menu creation at a minimum while simultaneously providing implicit pressure to maintain the status quo. The result has been the addition of many prepared and semiprepared foods, an occasional meal at a national chain restaurant, and an overall continuation of life as it has always been. Change has tended to come more at the behest of returning prodigals than the arrival of outsiders.

The creation of "historic" food experiences for urban foragers is a thriving source of income for rural churches and community organizations throughout the United States. These ladies are making apple butter in iron kettles over wood fires as a church fundraiser in West Virginia. (Richard Pillsbury)

The Northeast cuisine region thus is dominated by large areas of historic interaction zones throughout almost all of the eastern edge of the region. The foodways of these Industrial Age immigrants continue to dominate large parts of the five largest cities and several of the smaller ones. The suburban fringes of all larger urban areas, including virtually all of eastern Massachusetts, southern Connecticut, northern and central New Jersey, southeastern Pennsylvania, the Baltimore-Washington corridor, and suburban Washington stretching southward to Fredericksburg are dominated by national trends leavened by the continuing effect of the late-nineteenth-century migrations. The traditional fabric appears to be almost everywhere but actually still dominates only in those areas that have had little economic growth and in-migration over the past century. Central Pennsylvania and New York, northern Maine, and scattered sections of northern New England—those few places where weekenders don't dominate the scene—still maintain the old ways, though it is becoming more and more difficult to differentiate until the first bite between the continuing scene and the artfully created artifact.

The Midwest

The cuisine of the Midwest is mostly a product of the Industrial Revolution. Settlement of the rural Midwest expanded rapidly in the 1820s after the completion of the Erie Canal. The northern areas were initially settled by New Englanders; the central area was primarily populated by westward-moving people from the Middle Atlantic. The areas along the Ohio River are almost as southern in character as the southern core region itself.

Germans settled in virtually every area of the rural Midwest after 1820, becoming the dominant ethnic group in much of a wide band spreading west and northward from Ohio across the central corn belt to southern Wisconsin and Minnesota. Cheap lands in the grasslands carried them ever westward until the water gave out and a viable agriculture became impossible. Joining the Germans after 1840 was an ever-increasing stream of Scandinavians driven from their farms by the agricultural recession sweeping northern Europe. Settling initially in Illinois, Swedish, Norwegian, Finnish, and even a few Icelandic farmers soon followed the Germans across Wisconsin and Minnesota and the American grasslands. Though *My Ántonia* by Willa Cather may be the better known account of their struggles, Ole Rölvaag's *Giants in the Earth* paints a far more poignant portrait of their adjustments to this new life.

A broad ethnic and cultural disparity developed between the rural and urban Midwest during the late nineteenth century. The cities received immigrants from much the same areas of Europe, but these immigrants had few cash resources and were forced into wage labor in the growing cities. Jammed into ethnic ghettos, these immigrants too faced greater cultural intermixing and a reliance on purchased foods that may or may not have been what they wanted. Later these same cities became the homes of many of the descendants of those farmers, but the anonymity of urban life made it difficult to maintain their traditional ways as an integral part of their daily lives.

St. Louis, Milwaukee, and Cincinnati developed strong Germanic biases in their populations; Detroit, Minneapolis–St. Paul, and Chicago evolved with much greater mixtures of industrial-age immigrants. All have large populations from the Upland and Lowland South today. Like the eastern megalopolis, these cities continue to have strong industrial-age components in their home cuisines; in the suburbs more nationalization has taken place. Minneapolis–St. Paul also has a very strong Scandinavian heritage today; the children from nearby farms have moved into the cities, seeking shelter from the harsh economic environment of the upper Midwest.

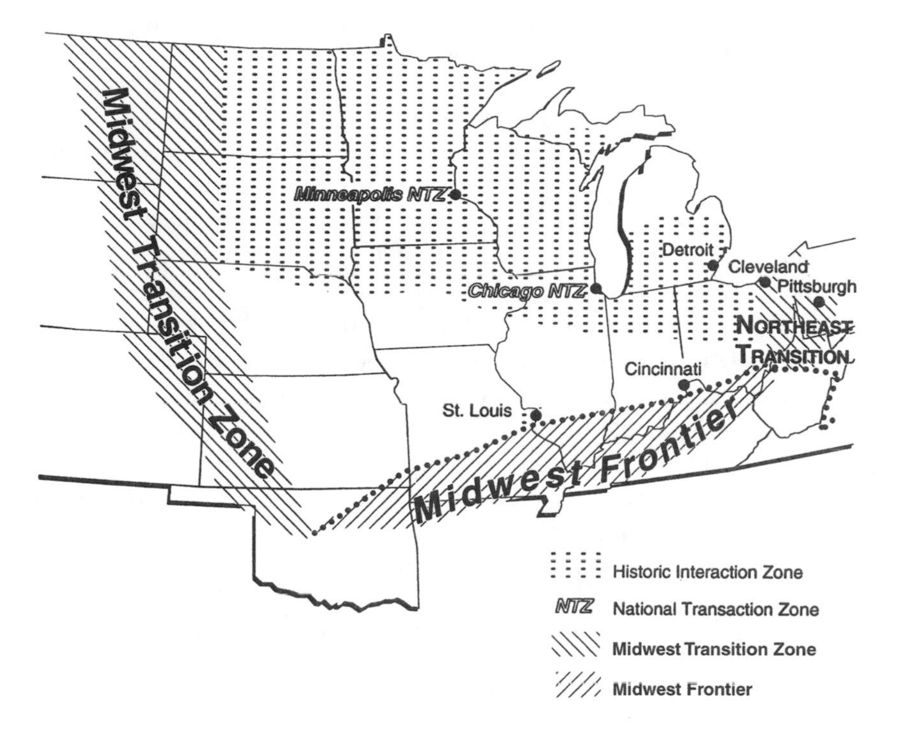

Contemporary Diet Regions: The Midwest

The geography of contemporary cuisines in the Midwest thus is much simpler than that found in the Northeast, and larger areas of historic interaction are still extant across most rural areas. The core of the region, both rural and urban, was so deeply affected by the arrival of millions of Europeans in the late nineteenth and early twentieth centuries that broad regions of historic interaction stretch across the heart of the Midwest and westward onto the American grasslands. The urban cores have also been almost totally transformed, first by the arrival of millions of southerners of both European and African American heritage and in recent years by just as large numbers of Hispanic and other new immigrants. All of the large cities and many of the smaller ones are surrounded by endless miles of suburban growth; these suburbs bring in and are dominated by national cuisine trends and patterns. Finally, some small areas of the traditional South can be seen along the Ohio River and in the most southerly sections of the grasslands, but they have played a relatively small role in the evolution of this region.

The region's consumption patterns clearly reflect the conservative character of the population. Dairy consumption continues to be the highest in the nation, and this is the only area in which it continues to increase. The continued northern European influences are reflected in higher consumption rates of potatoes, many pastry products, and red meat. Consumption of fresh vegetables continues to be the lowest in the nation. This is a region in transition; rural depopulation is a fact of life in many areas. Boarded storefronts, empty schools, and farm sales and consolidations are all frequently seen in the rural Midwest, making it increasingly difficult to continue the cuisine and other traditions associated with the region.

Little of the ethnic cuisine usually associated with the Industrial Revolution, primarily from eastern and southern Europe, is found in the midwestern restaurant scene. One examination of the geography of ethnic restaurants, for example, found that Chinese food was the most important cuisine in much of the north-central region. This finding clearly speaks more to the paucity of ethnic restaurants than to the frequency of Chinese restaurants (Zelinsky, 1985). Restaurants featuring German foods have never been common in America, though German and Scandinavian specialties are the most typical in the rural Midwest. An interesting countercurrent to these generalizations has been the region's embrace of pizza. Though pizza consumption is not particularly higher than other urban areas in America, it is interesting that the three largest pizza restaurant chains and the three largest frozen pizza companies all call the Midwest home.

The South

The South continues to be the most continuous bastion of tradition in America. Most of the South has shrugged off the twentieth century, and traditional fare continues to dominate almost the entirety of its rural areas beyond the urban fringe and the expressway. Red meat consumption continues unabated. It is the only region with above-average pork consumption, and its high fish consumption stems from traditional regional preferences, not modern adjustments. Other indications of the continuing importance of traditional consumption patterns are the above-average use of plain flour (to make biscuits), fats and oils (to indulge the southern penchant for overcooked, greasy vegetables), sweet potatoes, sugars (for sweetened iced tea), canned vegetables, and soft drinks. These patterns are changing; for example, women who work outside the home are probably not willing to get up early and make biscuits, as their mothers did. Except for various cola products, all of the previously mentioned foods, and even grits, are not consumed as much as they once were. Southern foodways continue; they are just not practiced as intensely as before.

There are areas of change, of course, most notably in northern Virginia, along the Sea Island coast of the Carolinas and Georgia, in Florida, and in and around the region's growth cities. The encroachment of Washington, D.C., southward into the Virginia countryside has inevitably transformed more and more of that state into Yankee territory. Proud of their southern heritage, increasing numbers of northern Virginians are southern more by history than practice.

The Sea Island coast vacation and retirement retreat now extends as a series of almost continuously walled golf and beach communities from Cape Hatteras to northern Florida. Even larger numbers of in-migrants live outside the gates in compounds scattered throughout the piney woods and within easy driving distance of the region's outstanding golf courses and beaches.

Adopting a combination of national foodways and local favorites, this region has become a suburb out of control. The cuisine of Charleston, South Carolina, the largest urban center in the region, well reflects its role in this new environment. Touting itself as an island of civility and southern manners, the city offers a broad fare ranging from traditional southern to nouveau southern to bagels and lox to pasta as well as a full range of national franchise outlets. Savannah, though not as advanced down this path of nationalism, offers almost as wide a variety, including probably the finest single restaurant along the entire coast. Hilton Head, in contrast, is the ultimate nationalized hamburger alley.

Southern Florida has never been a part of the true South, and the development of the Orlando recreational complex has pushed this Yankee outpost northward to suppress ever more of the traditional sawgrass South, which once dominated northern Florida. South Florida began evolving as a winter resort after the completion of railroad connections with the North in the early twentieth century. Though known in the past primarily for its large numbers of Jewish and other residents from the Northeast, south Florida actually has long been home to large numbers of people from throughout the Midwest and Northeast with a variety of ethnic and religious backgrounds as well as increasing numbers of wintering Canadians. The cuisine that evolved as a result was pure diner food composed of large portions of traditional American favorites, late-nineteenth-century immigrant specialties (especially pasta), and local seafood.

The modern era has brought a virtual avalanche of permanent and temporary residents from the Caribbean to south Florida, most notably from Cuba but actually from throughout the islands and northern South America. Black beans, jerked-almost-anything, exquisite Cuban sandwiches, conch and other Caribbean seafood delicacies, brilliant spicing, and lighter cooking have brought a new vitality to the cooking of the affected parts of the region. Use of the black bean, seldom seen in the United States thirty years ago, has spread from this region to the most remote corners of the nation, especially to many restaurants targeting consumers in their twenties and thirties. In many ways the fusion of preexisting local favorites and Caribbean foods has given this region one of the most interestingly different and innovative cuisines in the nation, though this cuisine is little recognized as yet. All it lacks is a Paul Prudhomme to lead the way into the national spotlight.

"Sunbelt" has become almost synonymous with "economic growth" over the past twenty years. Atlanta, the undisputed capital of the new South, along with Miami, Charlotte, and Houston, serving their respective subregions, all can boast of superb examples of the new national cuisine dominated by the trends discussed previously. Although they call their city the capital of the new South (when they aren't calling it "the world's next great city"), most Atlantans are hard pressed to name ten independent "southern" restaurants—and virtually all of those are located in lower-income neighborhoods. Largely nationalized with some vestiges of their traditional roots remaining, smaller cities such as Nashville, Birmingham, Knoxville, and Raleigh-Durham still retain a southern character in their cuisine, though it too is fast disappearing.

New Orleans has always stood alone. Long perceived as a southern city, it never has been. Established in 1718, the city became a transaction center specializing in exporting cotton, tobacco, corn, salt pork, and a thousand other items from the interior while being the primary entrepôt for much of the Midwest and the central South. Its uniqueness is generally ascribed to its early French and Spanish heritage, but the city was actually home to a wide variety of immigrants who took advantage of cheap fares on empty freighters returning to New Orleans for cotton and Midwest grains. Large numbers of Germans and Italians settled in the city prior to and after the Civil War. Caribbean influences were strong from the beginning, partially because of the early importation of slaves from the Caribbean sugar plantations prior to the banishment of such practices in the early nineteenth century, and later as plantation owners found it to their advantage to move to the Mississippi delta with their changing political fortunes on the islands.

The result was the evolution of a rich cultural heritage. Canal Street became the "neutral zone" between the Creole and the non-Creole, but the search for good food brought an intermingling of cultures and foodways. The city's cuisine took on a character of its own—part Creole, part Cajun, part Caribbean, part European, and all very much American. Though a major transaction center, New Orleans has never been an American city in the same way as Dallas and Denver and Atlanta. Exotic, dirty, exciting, and devoted to good food and good times, New Orleans has been home to some of the nation's most famous restaurants ever since people started keeping track of such things.

Much of the South is on the brink of change; nonsoutherners pour into the region in ever-growing numbers. Factories are sprouting in large cities and small towns, providing greater disposable income and a steady stream of technicians from outside the region. The continued expansion of the broiler industry in Arkansas, northern Alabama and Georgia, and North Carolina is also attracting large numbers of immigrants; some are legal, some not, but all come ready to work and with their own cuisines and food preferences. Primarily Mexican but increasingly from Southeast and East Asia as well, these workers are finding steady employment in those jobs that locals are leaving as they find better work in the new manufacturing plants down the road. This influx has brought Mexican and Chinese restaurants in profusion to places that just a decade ago saw pizza as exotic. As social stresses build, so does the introduction of new foods and cuisines.

In many ways the cuisine regions of the South exemplify what such regions ought to be—straightforward with little overlap and easily fitting their definitions. The vast majority of the South is still comfortably dominated by the traditional fabric with much of the early diet still visible, if not dominant. There are only a few historic interaction zones, most notably along the western edge of Texas and south Florida. Finally, there are a number of fairly well defined national interaction zones, most notably surrounding Atlanta, New Orleans, Memphis, Nashville, Richmond, and the linear Metrolina urbanized corridor stretching from Durham (North Carolina) to Greenville (South Carolina). Quite predictably, the Sea Island coast of Carolina and Georgia is a nonurban zone of nationalization, as is virtually all of Florida south of Gainesville away from Miami.

The South fades as it moves westward to eventually disappear in the Texas hill country west of Austin. This was an area of diversity from the beginning. Germanic central European immigrants began moving into the area soon after 1830, farming where possible, raising livestock in the drier areas. The westward-moving cattle culture largely passed over this area because it was too dry, but some people did settle. The Amerindian population was small and quickly displaced, but Hispanos also occupied the fringes; and Mexican Americans more recently have drifted into the area as laborers and farmworkers. The result was a complex amalgam of foodway transitions and frontiers, and the region continues so even today.

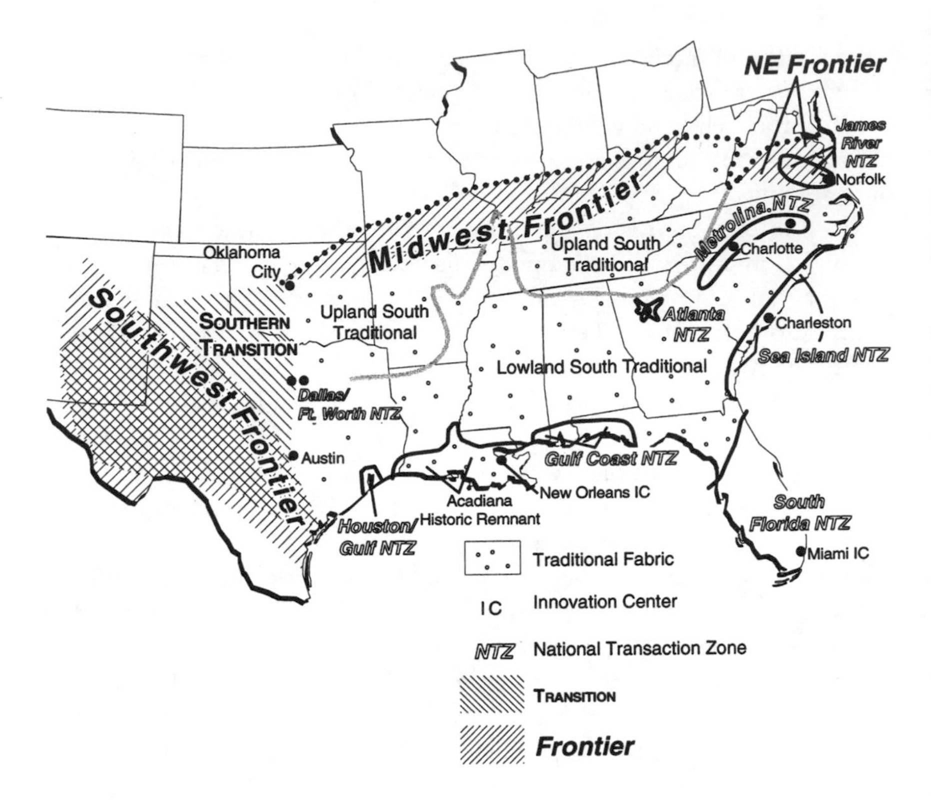

Contemporary Diet Regions: The South

The regional cuisine has come to carry elements of each of the major cultures with little fusion. The southern cattlemen who settled here learned to give up their pork, and barbecued meat today is either beef brisket or ribs. The barbecue equipment, however, is much like that seen even today throughout the South, as are many of the sauces. Pinto beans have replaced the black-eyed peas to the east, but they are still cooked much the same and are quite distinct from those of the pure Hispano areas to the south and the midwestern-influenced areas to the north. Chicken-fried steak sometimes seems to be the national dish of Texas and is seemingly represented everywhere in the region, including in the transition zone to New Mexico. Biscuits are available for breakfast in most Anglo restaurants west of Austin through southern New Mexico, although they begin disappearing quickly further west. The central European introductions of the potato and the cabbage, however, also play important roles in the regional diet. Coleslaw and potato salad are standard side items even in barbecue restaurants. Real hash browns, not those frozen, deep-fried quick-service abominations, are found everywhere. Finally, it is no longer necessary to go to a Mexican restaurant to purchase Mexican food. Burritos, carnitas, huevos rancheros, enchiladas, and rice are almost universally available in restaurants and are also eaten at least occasionally in even "Anglo" homes. Guacamole, chilis, and chorizo may still carry an ethnic connotation, but one would never know it by those who consume them regularly. This must be the jerky capital of the world with almost as many varieties available on convenience-store counters as smokeless tobacco tins in the Carolinas. Even today the vegetables consumed, however, tend to either be twentieth-century mainstream American or have a Hispano heritage.

The Southwest

Anglo cattle and traditional Hispano farm life permeates the Southwest, creating a region with many central themes and great diversity. Corn, beans, peppers, and a host of other Central American domesticates came early and still play an important part in the region's dietary character. Tortillas, a corn flat bread, was the staff of life; beans, small amounts of goat and pork, and a variety of condiments gave the dishes character. Several distinctive Hispano sets of foodways evolved across the region; the isolation of the Texas Rio Grande, New Mexico Rio Grande, California, and Arizona communities allowed each to develop independent identities.

The invasion of Americans and Europeans in the first half of the nineteenth century disturbed this established pattern. Westward-moving southerners brought their cattle culture into the region; it mixed with the established Spanish pattern to evolve into a distinctive American system prior to 1850. American traders and ranchers began settling across Arizona and New Mexico about the same time, establishing an independent area of dual lifestyles in those areas as well. Two separate sets of foodways long coexisted in the region; each was maintained primarily by the people who founded it, and remarkably little crossover occurred prior to the twentieth century. The Anglo cattle culture attempted to maintain its southern and a few central European food themes, accepting by necessity beef and pinto beans as replacements for what they had consumed before. Biscuits are available in most of the region, but so too are cornbread and tortillas.

Hispano cuisine has been overwhelmed by Anglo ingredients over the past century yet has maintained its essential character. The basic dishes—tamales, burritos, enchiladas, rice, beans, and the rest—remain at the core of the menu. The details of most of these dishes, however, has changed. The tortilla remains the staff of life, but with the invention of the tortilla machine in

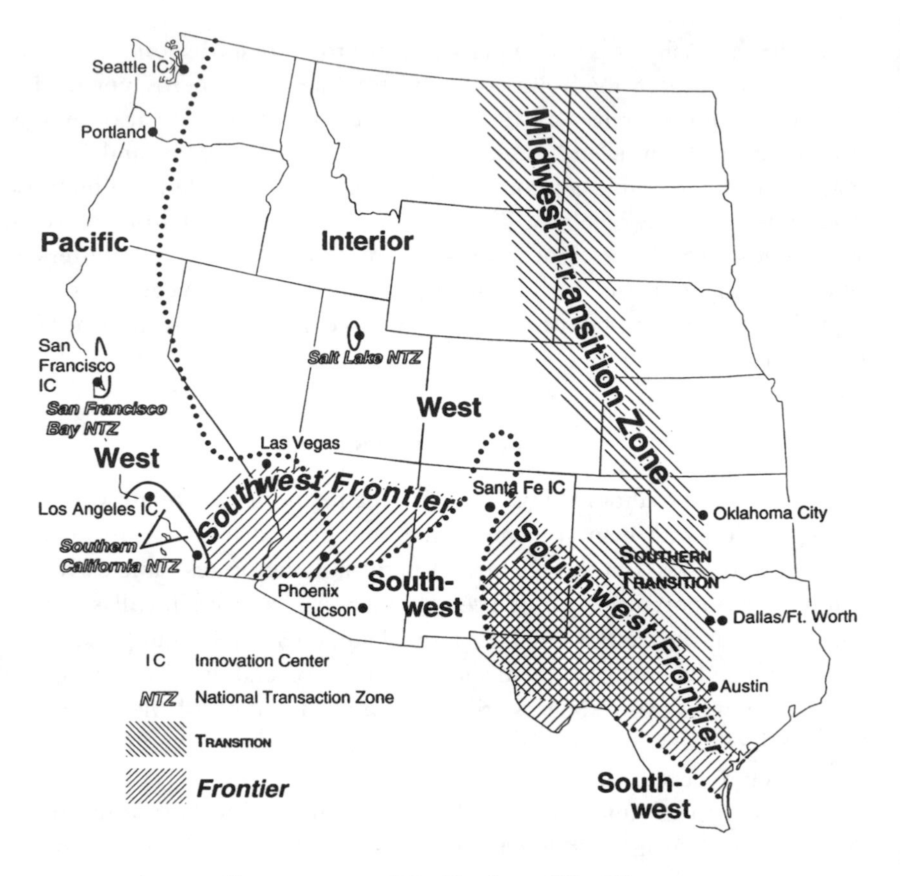

Contemporary Diet Regions: The West

Chicago in the late nineteenth century and the increased availability of wheat flour, this flat bread was more often than not made with wheat flour. Shredded meat is still the filling in most dishes, but goat and game have largely been replaced by pork, beef, and chicken. The introduction of goats by the Spanish brought goat cheese, which quickly became an important part of many Hispano dishes, though today traditional Anglo cheeses are more likely to be used in everyday cooking. Sauces too are little changed on the surface, though many are likely to be composed in part or completely of ingredients in processed form. Finally, rice, presumably introduced by the Spanish, has become an integral part of meals in almost all of the Southwest.

The invasion of large numbers of Mexican Americans and "Anglo" easterners into the Southwest since the 1960s has had a great effect on the foodways of the region. There has been a general lowering of the barriers between the two cuisines. Thus the menu of Ralph's Breakfast and Bar-be-cue, a very traditional Texas general restaurant, presents a breakfast menu dominated by traditional southern favorites, the fajita omelette and a breakfast burrito. The lunch and dinner menus concentrate on beef barbecue but also include fajitas, burritos, and a selection of what are actually fusion dishes.

Santa Fe and the upper Rio Grande valley have evolved into one of the most innovative cuisine zones in the nation, fusing traditional regional specialties and contemporary American cooking into new dishes and dining treats. Blue corn is possibly the signature ingredient, but the dishes are actually made unique by their blending of traditional ingredients and contemporary strategies. The evolution of a large artist and retirement community here has supported the creation of new restaurants and renewed interest in food.

Arizona and the associated California have been so overwhelmed by retiring midwesterners and general business growth that it is difficult to find the traditional among the national trends in many areas. Although lingering discrimination keeps the cuisines largely apart, the lines are becoming increasingly blurred among the younger residents, who eat large quantities of national foods.

The Interior West

The intermontane West remains as traditional in its outlook as anywhere in the country; the only problem is in defining traditional. This has always been a lightly settled area, the only dominant unifying force being the Church of the Latter Day Saints throughout Utah, southern Idaho, eastern Oregon, and much of Nevada. The Mormons began occupying this intermontane core of the interior West prior to the arrival of the cattle culture in the 1860s; thus this area never fully adopted the traditional "western" beef and beans cuisine that is generally thought to dominate most of the West. Instead, these primarily midwestern farmers created a diet that for all intents and purposes was a continuation of the midwestern milieu.

The early years were crowded with disaster and tribulations for the Mormons in their new home. Mormon farmers were encouraged to continue general farming as a form of protection from these early problems even after most other western farmers had turned to specialized crop and livestock cultivation. This incidentally gave the typical Mormon farm family access to a broader base of homegrown foods than enjoyed by most other western agriculturalists. The church also required its members to keep a large stock of preserved food on hand as a protection against and remembrance of those times of famine. This fostered the development of an important dehydrated-food industry, which in turn led to a much wider use of dried fruits and vegetables in day-to-day dining. Most farmers also continued to keep a few sheep to graze what was otherwise wasteland, a practice that was widely seen in the Midwest and Northeast until the early twentieth century. The remainder of the traditional interior West was forced to import most foods other than meat, which limited both the variety and amount of fresh fruits and vegetables consumed. Lamb and mutton are much more common in all of the intermontane regions than anywhere except in the Northeast because of the long tradition of shepherding in the drier areas. This more traditional diet continued until the introduction of hard-surface roads, though fresh produce consumption still continues at lower levels than national averages.

Westerners have always eaten in restaurants more than residents of any other region. Every town supported a café or two dispensing traditional "diner" food, but there were few communities that did not have a Basque hotel and a Chinese restaurant. Basques are a distinct culture group that seemingly originated in the French and Spanish Pyrenees. Basques have often been employed in the region as sheepherders, spending their summers with the sheep in the mountains and wintering in town or with the sheep in California. Hotels provided rooms for them when they came to town and family-style meals for both them and travelers. Dinners were especially geared to this trade and included heavy lamb stews, beans, and flan for dessert.

Most Chinese were driven out of the rural West by racial discrimination in the late nineteenth century, but a few remained in the intermontane mining towns. Most ultimately found work outside the mines, some starting restaurants that today are still found in many communities of any size in the region. These restaurants have become accepted in these communities, although the foods they serve have tended to be Americanized.

There have always been a few Mexican American mine and agricultural workers in the interior West, but in the post–World War II era, increased numbers settled north of Arizona and New Mexico for the first time. Drawn northward to take unfilled positions as miners and seasonal agricultural workers, more and more have settled permanently in these areas as local Anglo workers have migrated to the cities to find better-paying jobs. Bodegas and Mexican restaurants now flourish throughout the northern intermontane area, though it is unlikely that much Mexican food is cooked and served in most Anglo households.

Portions of the Rocky Mountains have become major four-season recreational areas, attracting visitors from throughout the world in the past few years. Food in Aspen, Steamboat Springs, and Park City (Utah) today has few links with the traditional West and largely reflects national and international preferences and trends.

The Pacific West

The Pacific West is the center of cuisine nationalization; indeed, many believe that this area is the center of all cultural innovation in America. The general consumption habits of the region reflect the basic trends in American foodways. Consumers here are abandoning red meat while embracing lower-fat chicken, turkey, and seafood products. Fat consumption is down over the past decade, and residents consume fewer sugars, fewer canned foods, and fewer soft drinks. Consumption of fresh fruits and vegetables is high and growing. All in all, the trends toward a healthier diet are everywhere.

The culture of California and much of the West Coast has long been schizophrenic, and this is reflected in its general dietary regime. The region embraces everything and anything that is new while continuing to honor traditional ways. Tens of thousands of legal and illegal aliens pour into California each year, adding their dietary elements to the already-crowded menu. The primary characteristic of the West Coast version of national cuisine is an overall tendency to grill as much food as possible. Fresh vegetables are preferred to cooked, and those that are cooked appear uncooked by national standards. Fusion cuisine, the blending of ethnic foods and preparation techniques, is an important undercurrent and probably explains the light sautéing of many vegetables. Pizza with pineapple and artichokes; stir fried anything; and description-defying combinations of fruits, exotic vegetables, and meats are all a part of this region's continuing search for the new and exotic while firmly hanging onto the familiar.

The West Coast has long been the center of restaurant cuisine innovation, and three separate innovation centers have evolved: Los Angeles, San Francisco, and Seattle. Los Angeles is the oldest and most well established. Its role in this field flowered in the 1960s, though obviously the McDonald brothers were local innovators from an earlier time. Many of the early restaurant chains had become spectacularly successful, and investors searching for new concepts haunted the Los Angeles area, where operators began creating concepts with the single goal of finding investors, going national, and gaining instant riches. This open atmosphere and growth economy supported the development of a large number of independent restaurants offering something that was different. Some of these concepts were abysmal, but some were exciting; chefs from throughout the nation came to look for new restaurant themes and entrée concepts.

San Francisco has always been a restaurant town with a sense of tradition. Its restaurants reflect this ethic, tending to focus on proven rather than exotic themes. In many ways the Fog City Diner is a good example; it brought traditional diner food—meat loaf, pot roast, and mashed potatoes—up to today's standards. Better than Mom's, the traditional dishes served in such restaurants are part of a new national retro restaurant industry. As one of the nation's most popular tourist destinations, San Francisco restaurant operators are guaranteed a constant stream of visitors ready to try new concepts, though most of the very innovative eateries are located in the area's less frequented neighborhoods and surrounding communities. Visitors in search of new concepts that don't venture much beyond Fisherman's Wharf and Union Square are doomed to disappointment.

Seattle is the newest and least well established restaurant innovation center on the West Coast. Taking advantage of the region's local fish supply and an increasing Pacific Rim presence and with a certain contempt of their neighbors to the south, the city's innovative restaurateurs have created a fusion cuisine that is setting its own pace. The nation's continuing reluctance to eat fish has tended to discourage Seattle restaurateurs from creating too-exotic fish entrées, but the city's other innovation—exotic coffees—has quickly been taken up by the rest of the nation. Coffee consumption, which dropped precipitously for several decades, has leveled off as Americans have embraced the coffeehouse and "designer coffees."

The rural West Coast is a bit less extreme in its foodways, though innovation does seem to be a part of the whole region's psyche. The home diet still looks quite traditional at first encounter but has undergone more change than anywhere else in the country. Fresh vegetable, fruit, wine, turkey, and wok sales all continue to increase in the most rural of areas. Soft-drink consumption is up here, as everywhere else, but the region has developed several "natural" brands to provide the illusion of nationalization without giving into it totally.

A Few Final Words

The regionalization of American cuisine has always been difficult because of the restless character of its population. Few places have maintained a stable population with little in- or out-migration over many generations to create stable ethnic homelands. Almost every American community is constantly absorbing change. The result is an exceedingly complex set of images reflecting the interplay between the past and the present, the individual and the general. Regional portraits can be little more than fleeting glimpses of rapidly changing ways of life.

Regional preferences and biases do continue, although in less defined terms than in the past. There was a time not long ago that the included drinks in every southern grill were tea (iced and sweetened, never hot), coffee, and buttermilk; a fried tenderloin sandwich was on every café menu in the Midwest; and the only grocers with piroghis in the frozen-food case were in the Northeast. These traditions have mostly faded, but that does not mean that regional preferences no longer exist. New patterns following new rules have taken their place.

11

Continuity and Transformation: Last Thoughts

Too busy to shop, too tired to cook.

—Dennis Leonard, 1996

Our dining preferences and habits are a microcosm of the complexity and changing character of American life. Each meal we consume reflects not only our societal and individual pasts, but the interplay of our current social milieu as well. The dinner described on the next page—a gathering of friends around a meal of traditional foods—well reflects this interplay of past and present, of individual and group dynamics. The food tradition was of southern Louisiana; yet none of the group had actually been born there. The tradition was authentic, however, for it represented the past of those who prepared it; the food was authentic because it was based on recipes handed down through the generations, not learned from a book.

It has been said that the only constant in American life is change. The time when individuals ate much the same foods as their parents has passed, just as the tendency to live in the same community and seek similar employment has passed. But within this changing scene is also a sense of continuity. Life in the suburban scene described here is a microcosm of that battle between continuity and transformation. Old foods and new friends; old friends and new foods—these are themes that pass through our modern life. The past is carried forward in our memories and melded with the present and eventually carried into the future by our children. Yet although Sean and Sedric may enjoy their grandmother's fried catfish and sweet potato pie, they will never understand its importance regarding who she is. Nor will she totally understand the importance to them of eating a Domino's pizza while watching a televised football game in the recreation room when they are adults.

The American diet has been transformed over the past 400 years. Little is left of the original colonial European diet. Potages, salt pork, and rusk breads have passed into our collective memory and have been replaced by stir-fried vegetables and grilled, farm-raised salmon. The transformation has taken place partially because of new technologies, partially because of new lifestyles, and partially because of the millions of immigrants who have come to these shores over the years. The American culture has always had a huge appetite for new things and new people but simultaneously has maintained an overriding chauvinism and sense of superiority over them. The larger process has always been toward amalgamation, not coexistence.

That does not mean that everyone dresses alike or talks alike or eats the same food. As surely as there has been a nationalization of our way of life,

Dinner with Old Friends in the New South

It was getting toward dusk as we turned the corner and drove down the street among the faux European–style two-story suburban houses. Windward is your typical 1970s suburban multiuse cocoon—housing, office space, light industrial, hospital, a new regional mall down the road. Faltering for a decade, the development began growing in the 1980s as the city grew out to meet it. The $300,000 price tags on the homes ensured that they were all different; yet with their common age, size, and fake stucco exteriors, they all looked much the same.

We checked the cars as we pulled into the driveway: Our hosts' cars—Stacey's black S-Class Benz and Kellie's baby 300 E—were in the garage. Stacey's Jeep with the SK Landscaping sign was out in the driveway—most of his customers couldn't deal with the fact that their yardman drove a Mercedes sedan (and was probably worth more than they were). Only a single unidentified Mazda sat out on the street. It was going to be a quiet dinner. We had expected a bigger crowd.

Mama, Kellie's mother, was cooking her last meal before heading back home to the San Francisco Bay area on Monday. She was cooking an old-fashioned meal for us because Kellie and I had gotten into a discussion about the food she ate while growing up in the San Francisco Bay area. Mama, who actually is younger than I am, had grown up in Austin, but Kellie's dad was from New Orleans. The food that Kellie remembered best was Creole. Tonight was to be a southern Louisiana night with a little Mississippi delta (Stacey's family was from Mississippi) thrown in for good measure.

Stepping into the kitchen we were not disappointed about the crowd. There might have been only the one car, but a friend of Mama's from California was sitting at the kitchen table, and Kim, the sort-of baby-sitter, was with her two-year-old son, running around underfoot with Kiara, Kellie's two-year-old. Mama was behind the peninsula with mounds of pots throwing out smells of greens and yams and who knows what. Patricia and Mama and Kellie were soon shrieking with laughter over Mama's account of their expedition downtown to Auburn Avenue the previous week. I corraled nine-year-old Sedric long enough to discover that Stacey was downstairs. I assumed he was watching the National League playoffs, so I moseyed down to kill a little time. The game was on, but the Yankees were playing. No one in Atlanta cared much about the Yankees. Stacey was at the computer, working on invoices, and we had hardly gotten into his modem problems when Sedric arrived to tell us it was time for dinner.

It was a classic family dinner. Four kids swarmed around, afraid they weren't going to get enough to eat even though the peninsula was covered with food.

(continues)

(continued)

Sean didn't feel well. He filled his plate and hightailed his twelve-year-old self into the dining room out of the line of fire. The two little ones got their plates and sat on the stools at the counter—they weren't ready to stray from the action. Sedric, the quiet, thoughtful one, stood to one side, watching everything going on, though I noticed he soon had a plateful. He didn't follow his brother but stayed in the room, watching the constant interplay.

I finally started down the line of bowls and platters—catfish strips fried in cornmeal, fried chicken, black-eyed peas, rice, and potato salad (a little Texas influence from Mama's growing up in Austin) took up the counter. Kim was over to the side popping corn muffins from a tin. A giant bowl of collards appeared on the counter.

"You use peppers?" Stacey asked as he poured some sauce on his greens and offered the bottle to me.

"Can you eat greens without peppers?" was my only reply as I poured on a liberal dose and tasted. The sweetish flavor of properly cooked fresh collards floated through my mouth.

"This sauce is a bit hotter than mine," I commented as I picked up the bottle again and checked its origin—New Iberia, Louisiana, but a hotter brand than the one I used. I made a mental note.

We settled around the kitchen, and Mama and I began talking about when she left the South and I came to it in the early 1960s. "I was in the second underground class at UT," she said. Patricia looked a bit confused (not having lived in the South during school segregation), but I knew what she meant, having gone to school at LSU at the same time. "But I couldn't keep my mouth shut and my mother decided that I ought to go live with my aunt. I transferred to San Francisco State to finish my college."

"Was it any better?"

"Some."

Fearful of the dinner getting cold, I turned to the food. I had already taken teasing over the catfish because everyone knew that I don't eat catfish, but I took a tentative bite anyway. The fish melted in my mouth. "How did you cook this?" I interrupted, knowing that in this crowd no conversation lasted more than three sentences without redirection anyway.

"Sprinkled on some Creole seasoning, a little salt, deep fried it."

"Dredge it in cornmeal?"

"Of course."

"Where did you get the seasonings?"

(continues)

(continued)

"Kroger. I could make my own, but I didn't have time."

"They don't cook like this in Austin."

"No, but my husband's family did. I had to learn from them. That man couldn't have a dinner without rice."

"Speaking of rice, this isn't traditional rice."

"Oh, Richard, I cooked that," Kellie interrupted.

"But you don't cook rice, Kellie. This from a bag?"

"You know it." Laughter erupted on our side of the room. Everyone understood that between taking care of the kids and operating her thriving real estate business, Kellie didn't cook from scratch unless there was no other way.

"Hey, Mama" suddenly exploded from across the room, and everything ground to a halt as Tony strolled in from outside with his girlfriend—a pretty young woman from Chicago. He went over to Stacey while she filled her plate and joined our conversation, which had now wandered into why men were such rats.

This was a good time for me to leave the ladies sitting around the table. I sauntered over to the desserts—banana cream pudding covered with a mound of meringue (not old-fashioned but certainly an indelible part of southern cooking as long as I had lived here), fresh peach cobbler (none of that canned stuff that so often parades as cobbler in the typical southern restaurant these days), and sweet potato pie. I passed on the pie—I hate sweet potatoes—but then reconsidered, thinking about the catfish. Maybe I had better cover my bets. I took a small piece.

Dropping into the male conversation, I realized that I was out of my league in their discussion of the upcoming football season, so I drifted back to the women, hoping they had passed through the rat conversation. Sitting again at the table, I took abuse over the pie.

"You know that's sweet potato pie," Patricia opined, taking a sample off the end and then reaching for another.

I took a bite; it was heaven. I grabbed the plate, saying, "There's a whole pie over there; get your own."

She ignored me and took another "taste" and then asked for the recipe.

Mama looked a bit embarrassed and then said, "I don't have a recipe. I know what goes into it, but I don't have any idea how much. I just cook it the way my grandmother taught me."

The two women got out a napkin and started working out the recipe. With ten years of college between them, they ought to have been able to handle a

(continues)

(continued)

small problem like this. After a few minutes I heard Patricia say, "It's pretty much like mine, just an extra egg to make it lighter."

"And a half stick of butter," Mama added belatedly.

"I guess that's why I like it," I commented. "It's lighter than what I am used to. I still remember my first fall in Louisiana. I grabbed a piece of pumpkin pie at the university cafeteria and almost choked."

"You just don't appreciate good things," my sweet-potato-loving wife commented.

I decided it was time to move again and dropped into the conversation between Stacey and Tony and some fellow who had slipped in when I wasn't paying attention. His plate was piled high and their conversation had moved to laughing about their days of college football in the Bay area and the good times before they scattered across the country to eventually reunite in Atlanta.

And before I knew it the dishes were in the dishwasher and we were saying our fond good-byes out in the driveway. We took away some warm memories and a doggie bag of sweet potato pie.

there has simultaneously been a continuing centripetal motion ensuring the maintenance of social and geographic separateness. The national cuisine and distinctive regional cuisines of the past may have disappeared, but they have not been replaced by homogeneity.

Yesterday I went shopping for a gift that was so unique that I didn't even know where to start the search. My first stop was at a suburban shopping mall filled with conventional upper-middle-income housewives and their children shopping for Christmas. Failing to find what I wanted there, I moved on to an inner-city "counterculture" shopping district filled with skinheads, artist wanna-bes, suburban teenagers trying to blend, and shops featuring everything from futons to T-shirts proclaiming "Die Yuppie Scum." Failing again, I proceeded to a bustling, aging shopping center where I had once seen an item similar to what I was seeking. I entered the mall to the sound of a gospel concert in the atrium and hundreds of African American shoppers clapping, singing along, and taking a little time out of their busy day to enjoy the moment. The store I sought was no longer there, so after tarrying too long to listen to a group of young girls who sounded like they had just arrived from heaven and after purchasing a homemade tape from their "manager," I successfully finished the day at a flea market–world bazaar in a retrofitted Kmart in the heart of the immigrant district. Each of these shopping environments was filled with a distinctive set of determined shoppers, bored husbands, running children, harried shopkeepers, and diverse shops. The trip had been like traveling to four different nations; yet they were all within a few miles of each other. Regionality clearly is not dead.

Culture-based regions continue to exist, but they have taken on a new form. The concept of "culture regions" covering vast miles of forest, desert, or farmland is no longer relevant in this urban society. That does not mean that there are no distinctive American subcultures or that those subcultures do not have a geographic expression. What has changed is the way in which lifestyle groups agglomerate. The regions that express cultural idiosyncrasies too have taken on new morphologies. The new regions are more likely to be discontinuous exclaves, appearing independently in each of the nation's growth cities; each is a manifestation of the same forces, so that in the final analysis they look much alike. The physical expression of these places in any individual city may also be discontinuous. These places are most visible where the members meet—shopping districts, entertainment zones, and the like—and weakest where the people actually live because residential location today is more likely related to job access and income than to lifestyle. Residual cultural homelands of another time still inhabit vast sweeps of rural America, but the city belongs to the present.

Much of this apparent randomness of the geography of American foodways and lifestyles does not stem from a lack of order but rather from the absence of an effective perspective from which to view it. There is continuity and there is order in the seemingly ceaseless changes taking place around us. We may say that the nation's foodways reflect a pervasive ennui—that the restlessness is little more than the physical expression of a lack of a sense of place, both figuratively and literally. We may lament that we have lost a more meaningful past and bemoan today's lack of character of place. Yet we know that one day those in the future will look back at our times and lament their passing.

Clearly most Americans know little of their roots, where they fit into society, or where that society is going. This placelessness has created a restless society constantly searching for a sense of identity and roots. Food has always been a major element in that sense of identity. When all else failed, one could go home to Grandmother's at Thanksgiving or Christmas for a repast of comfort foods or even create a meal at home that reminded one of childhood and renewed one's sense of tradition and place. Unfortunately, for all too many of us contemporary Americans, Grandmother now lives in a home, our parents have retired to Florida, and Little League and other activities leave little time for learning about family traditions. A dinner of Lean Cuisine in an impersonal suburban subdivision just doesn't give one a sense of place, attachment, or belonging.

As a society it has been difficult to accept that the past cannot be resurrected. The sense of placelessness that pervades suburban life has convinced many to attempt to re-create places and traditions that may or may not have existed. Regional foods and foodways have become big business as entrepreneurs attempt to fill the haunting emptiness felt by millions of Americans. Through all of this, and surrounded by hundreds of festivals and authentically re-created small towns celebrating important days that never happened, there remains the nagging question: Are we so busy trying to maintain the past because it is important to us, or are we just afraid of the void of the future?

A seemingly benign loss, this sense of place, but slowly we are beginning to realize that the place identity that was once an integral part of each of us provided us with a sense of security. It told us who we were and where we fit. As our world has become filled with uniformity—of buildings, homes, businesses, signs, and even species of grass in our ubiquitous lawns, it is becoming difficult to tell if we are residing in Long Beach or Hoboken. The importance of this to our self-image is revealed in the increasing attempts to "create" local images, festivals, and foods to make "our place" unique from all others. But even the ways that we do this are all the same. How many times have we awakened in a motel room disoriented and unable to remember where we were because our room looked like every other motel room we had ever been in? More important, if all destinations are alike, just like a place down the street from our homes, what point is there in visiting them in the first place? In the past we used food to help create a sense of place. Clambakes on Cape Cod, crawfish boils in Louisiana, and gatherings with oysters along the Chesapeake were all a part of those regional identities, but in this day of prepackaged foods and prepackaged places, we can bake a clam in Peoria and dine on crawfish bisque in South Carolina. In our drive to create place, we have destroyed it. David Lowenthal, a particularly astute geographer, commented many years ago that many Americans would rather visit the French Quarter at Disney World than travel to New Orleans—after all, the real place is dirty and reeks of stale beer and rot.

Ultimately this book is not really about food. This exploration of our changing diet has only casually been about how what we eat has changed over the centuries. The underlying goal from the opening "I don't eat no foreign food" to the closing "Too busy to shop, too tired to cook" has been to explore our changing lifestyles and self-image as reflected in the food that we eat. On this journey of continuity and transformation we began with a monotonous diet of salt pork and cornbread and traveled through technological, demographic, and intellectual revolution until our cuisine became a veritable world bazaar marked by exotic dishes and ingredients. During this trip our society has moved from a time when virtually every waking hour was devoted to finding and preparing food to a time when the frenetic pace of dual-worker households, long commutes, traffic jams, and trying to be too many things to too many people have brought us individually and societally to the point that we indeed are "too busy to shop, too tired to cook." As we head to a restaurant or home-meal-replacement emporium, we think only of a little body food to carry us through, but as we select from jerked chicken and stir-fry and fajitas, even the most unperceptive outside observer would know immediately that in contemporary American society, there truly are no foreign foods.

Select Bibliography

My interest in diet was first tweaked by reading Frederick Simoons's *Eat Not This Flesh*, an exploration of the origins and spread of food avoidances. This book suggested to me that a geography of diet might be possible; Sam Hilliard's *Hog Meat and Hoecake* showed one way that it might be done. Sam has been a great friend over the years, initiating me into the mysteries of the South, and the evenings with him and Jerry Holder were important in adding to my understanding of the South in particular and regionality in general. Becoming interested in urban diets in the 1980s, I wrote *From Boarding House to Bistro*, a geography of restaurants, which in turn led me to Philip Langdon's *Orange Roofs, Golden Arches*, a fascinating history of the modern American restaurant.

Harvey Levenstein's *Revolution at the Table: The Transformation of the American Diet* provided a starting point in the research for this book. No one can possibly write about the history of the American diet without reference to the works of Jones, Root, and, Tannahill, and Lowenstein's history of the American cookbook gives far more insight into the concepts of the period than a mere list of cookbooks. Similarly, no work on the geography of American culture would be complete without a perusal of the works of Wilbur Zelinsky and the always thoughtful J. B. Jackson.

Finally, I would like to point out two cookbooks that were especially important in bringing me a new understanding of American cooking. Karen Hess has written and edited several works, but her *Carolina Rice Kitchen: The African Connection* should be on the bookshelf of everyone who is interested in the regionality of American food. Purchased originally to learn more about my new home on Folly Island, South Carolina, it turned out to be one of the finest books on the evolution of American regional foodways ever written. Carrie and Felicia Young's *Prairie Cooks: Glorified Rice, Three Day Buns, and Other Reminiscences* is a cookbook of a different sort. The authors' charming reminiscences of their childhood in North Dakota bring the reader a deeper understanding of a regional diet from a humanistic perspective and of the modernization process generally. There should be more like this.

Achaya, K. T. 1994. *Indian Food: A Historical Companion.* Delhi: Oxford University Press.

Adelman, M. A. 1959. *A&P: A Study in Price-Cost Behavior and Public Policy.* Cambridge: Harvard University Press.

Allen, James P., and Eugene J. Turner. 1988. *We the People: Atlas of American Ethnic Diversity*. New York: Macmillan.

Anderson, Oscar E., Jr. 1953. *Refrigeration in America: A History of a New Technology and Its Impact*. Princeton: Princeton University Press for the University of Cincinnati.

Arnold, Pauline, and Percival White. 1959. *Food: America's Biggest Business*. New York: Holiday House.

Ayensu, Dinah. 1972. *The Art of West African Cooking*. New York: n.p.

Bard, Bernard. 1968. *The School Lunchroom: Time of Trial*. New York: John Wiley.

Barer-Stein, Thelma. 1981. *You Eat What You Are: A Study in Ethnic Food Traditions*. Toronto: McClelland and Stewart.

The Bedingfield Inn Cookbook. 1966. Lumpkin, GA: Stewart Count Historical Commission.

Beecher, Catherine. 1841. *A Treatise on Domestic Economy, for the Use of Young Ladies at Home, and at School*. Boston: Marsh, Capen, Lyon, and Webb.

Bell, Martin L. 1962. *A Portrait of Progress: A Business History of Pet Milk Company from 1885 to 1960*. St. Louis, MO: Pet Milk Company.

Bradford, William, and Edward Winslow. 1969. *Mourt's Relation, or, Journal of the Plantation at Plymouth*. New York: Garrett Press.

Brown, Linda Keller, and Kay Mussell, eds. 1984. *Ethnic and Regional Foodways in the United States: The Performance of Group Identity*. Knoxville: University of Tennessee Press.

Bruce, Scott, and Bill Crawford. 1995. *Cerealizing America: The Unsweetened Story of American Breakfast Cereal*. Boston: Faber and Faber.

Cahn, William. 1969. *Out of the Cracker Barrel: The Nabisco Story from Animal Crackers to Zuzus*. New York: Simon and Schuster.

The Canned Food Reference Manual. 1939. New York: American Can Company.

Carawan, Guy, and Candie Carawan. 1989. *Ain't You Got a Right to the Tree of Life?: The People of Johns Island, South Carolina—Their Faces, Their Words, and Their Songs*. Revised and expanded edition. Athens, GA: University of Georgia Press.

Carter, Charles. 1730. *The Complete Practical Cook: Or, a New System of the Whole Art and Mystery of Cookery*. Facsimile by London: Prospect Books, 1984.

Charleston Receipts. 1950. Charleston: Junior League of Charleston.

The Chicago Packer. 1915. Chicago: n.p.

Child, Lydia. 1836. *American Frugal Housewife*. Reprint of 1836 edition by New York: Harper & Row, 1970.

Claiborne, Craig. 1979. *The New York Times Cookbook*. New York: The Times.

Clark, Victor. 1916. *History of Manufactures in the United States, 1607–1914*. Two volumes. Washington, DC: Carnegie Institution.

Coe, Michael D., and Sophie D. Coe. 1984. "Mid–Eighteenth Century Food and Drink on the Massachusetts Frontier." In Peter Benes, ed., *Foodways in the Northeast, The Dublin Seminar for New England Folklife: Annual Proceedings. 1982*. Boston: Boston University.

Coe, Sophie D. 1994. *America's First Cuisines*. Austin: University of Texas Press.

Collins, Douglas. 1994. *America's Favorite Food: The Story of Campbell Soup Company*. New York: Harry N. Abrams.

Cooking with Condensed Soups. 1952. Camden, NJ: Campbell Soup Company.

Cooper, James. 1958. *The Last of the Mochicans*. Boston: Houghton Mifflin.

Cosman, Madeleine Pelner. 1976. *Fabulous Feasts: Medieval Cookery and Ceremony*. New York: George Braziller.

Cramer, Esther R. 1973. *The Alpha Beta Story: An Illustrated History of a Leading Western Food Retailer*. La Habre, CA: Alpha Beta Acme Markets.

Cummings, Richard O. 1970. *The American and His Food*. New York: Arno Press & New York Times.

Cunningham, Marion. 1996. *The Fannie Farmer Cookbook*. Thirteenth edition. New York: Knopf.

Davis, Pearce. 1949. *The Development of the American Glass Industry*. Cambridge: Harvard University Press.

Dent, Huntley. 1985. *The Feast of Santa Fe: Cooking in the American Southwest*. New York: Simon and Schuster.

Directory of Supermarket, Grocery & Convenience Store Chains, various years. Tampa, FL: Business Guides.

DuSablon, Mary Anna. 1994. *America's Collectible Cookbooks: The History, the Politics, the Recipes*. Athens: Ohio University Press.

Eames, Alfred W., and R. G. Landis. 1974. *The Business of Feeding People: The Story of the Del Monte Company*. New York: Newcomen Society of the United States.

Edgerton, John. 1987. *Southern Food: At Home, On the Road, in History*. New York: Alfred A. Knopf.

Eleventh Census of the United States, Manufacturing. 1893. Washington, DC: Government Printing Office.

Eustis, Célestine. 1903. *Cooking in Old Créole Days: La Cuisine Créole à l'Usage des Petits Ménages*. New York: R. H. Russell.

Farmer, Fannie. 1942. *The Boston Cooking-School Cook Book*. Seventh edition completely revised by Wilma Lord Perkins. Boston: Little, Brown.

Fenton, Alexander, and Eszter Kisban, eds. 1986. *Food in Change: Eating Habits from the Middle Ages to the Present Day*. Dundee: John Donald.

Ferguson, James L. 1985. *General Foods Corporation: A Chronicle of Consumer Satisfaction*. New York: Newcomen Society of the United States.

Fields, Mamie, and Karen Fields. 1983. *Lemon Swamp and Other Places: A Carolina Memoir*. New York: Free Press.

Flexner, Marion. 1949. *Out of Kentucky Kitchens*. New York: American Legacy Press.

Fowler, Bertrand. 1952. *Men, Meat and Miracles*. New York: Julian Messner.

Frederick, J. George. 1971. *Pennsylvania Dutch Cookbook*. New York: Dover.

Garrett, Elisabeth. 1989. *At Home: The American Family, 1750–1870*. New York: Harry N. Abrams.

Gazel, Neil R. 1990. *Beatrice: From Buildup to Breakup*. Urbana: University of Illinois Press.

Good Housekeeping's Book of Menus, Recipes, and Household Discoveries. 1922. New York: Good Housekeeping.

Goodrum, Charles, and Helen Dalrymple. 1990. *Advertising in America: The First 200 Years*. New York: Harry N. Abrams.

Grigson, Jane. 1983. *Jane Grigson's Book of European Cookery*. New York: Atheneum.

Grover, Kathryn, ed. 1987. *Dining in America, 1850–1900*. Amherst: University of Massachusetts Press, and Rochester, NY: Margaret Strong Museum.

Hampe, Edward C., Jr., and Merle Wittenberg. 1964. *The Lifeline of America: Development of the Food Industry*. New York: McGraw-Hill.

Hardyment, Chistina. 1988. *From Mangle to Microwave: The Mechanization of Household Work*. Cambridge: Polity Press.

Harrigan, Elizabeth. 1983. *Charleston Recollections and Receipts: The Recipes of Rose P. Ravenel*. Columbia: University of South Carolina Press.

Heller, Edna Eby. 1968. *The Art of Pennsylvania Dutch Cooking*. New York: Galahad Books.

Henisch, Bridget Ann. 1976. *Fast and Feast: Food in Medieval Society*. University Park: Pennsylvania State University Press.

Hess, Karen. 1981. *Martha Washington's Booke of Cookery, Transcribed with Historical Notes and Copious Annotations*. New York: Columbia University Press.

______. 1992. *The Carolina Rice Kitchen: The African Connection*. Columbia: University of South Carolina Press.

Hewitt, Jean. 1977. *The New York Times New England Heritage Cookbook*. New York: G. P. Putnam's Sons.

Hieatt, Constance, Brenda Hosington, and Sharon Butler. 1996. *Pleyn Delit: Mevieval Cookery for Modern Cooks*. Second edition. Toronto: University of Toronto Press.

Hill, Annabella. 1872. *Mrs. Hill's Southern Practical Cookery and Receipt Book*. Facsimile of 1872 edition by Columbia: University of South Carolina Press.

Hilliard, Sam. 1972. *Hog Meat and Hoecake: Food Supply in the Old South, 1840–1860*. Carbondale: Southern Illinois University Press.

Hochstein, Peter, and Sandy Hoffman. 1981. *Up from Selzer: A Handy Guide to Four Generations of Jews in America*. New York: Workman.

Hugill, Peter J. 1993. *World Trade Since 1431: Geography, Technology, and Capitalism*. Baltimore: Johns Hopkins University Press.

Humphrey, Theodore, and Lin T. Humphrey. 1988. *"We Gather Together": Food Festival in American Life*. Ann Arbor, MI: UMI Research Press.

Jackson, J. B. 1970. "Stranger's Path," reprinted in E. H. Zube, *Landscapes: Selected Writings of J. B. Jackson*. Amherst: University of Massachusetts Press, pp. 92–106.

Johnston, James P. 1977. *A Hundred Years of Eating: Food, Drink and the Daily Diet in Britain Since the Late Nineteenth Century*. Dublin: Gill and Macmillan.

Jones, Evan. 1981. *American Food: The Gastronomic Story*. New York: Vintage Books.
Kander, Mrs. Simon, and Mrs. Henry Schoenfeld. 1903. *The Settlement Cookbook, 1903: The Way to a Man's Heart*. Facsimile edition published by New York: Hugh Lauter Levin Associates, 1984.
Keir, Malcolm. 1928. *Manufacturing*. New York: Ronald Press.
Kitchen Kappers II. 1978. Chicago: Edison Park Lutheran Church.
Kittler, Pamela G., and Kathryn Sucher. 1989. *Food and Culture in America: A Nutritional Handbook*. New York: Van Nostrand Reinhold.
Klees, Fredric. 1951. *The Pennsylvania Dutch*. New York: Macmillan.
Kreidberg, Marjorie. 1975. *Food on the Frontier: Minnesota Cooking from 1850 to 1900 with Selected Recipes*. St. Paul: Minnesota Historical Society Press.
Langdon, Philip. 1986. *Orange Roofs, Golden Arches*. New York: Alfred A. Knopf.
Leonard, Dennis. 1996. Unpublished comments, Restaurant Finance Corporation Seventh Annual Conference.
Levenstein, Harvey A. 1988. *Revolution at the Table: The Transformation of the American Diet*. New York: Oxford University Press.
Linck, Ernestine S., and Joyce G. Roach. 1989. *Eats: A Folk History of Texas Foods*. Fort Worth: Texas Christian University Press.
Lord, Isabel, ed. 1924. *Everybody's Cook Book: A Comprehensive Manual of Home Cookery*. New York: Harcourt Brace.
Lowenstein, Eleanor. 1972. *Bibliography of American Cookery Books, 1742–1860*. Worcester, MA: American Antiquarian Society.
Mariani, John. 1983. *The Dictionary of American Food and Drink*. New York: Ticknor & Fields.
Martin, Edgar W. 1942. *The Standard of Living in 1860*. Chicago: University of Chicago Press.
McGaw, Judith, ed. 1994. *Early American Technology: Making and Doing Things from the Colonial Era to 1850*. Chapel Hill: University of North Carolina Press for the Institute of Early American History and Culture.
McMahon, Sarah. 1985. "A Comfortable Subsistence: The Changing Composition of Diet in Rural New England, 1620–1840." *William and Mary Quarterly* 42, 3d ser., pp. 26–65.
______. 1989. "'All Things in their Proper Season': Seasonal Rhythms of Diet in Nineteenth Century New England." *Agricultural History* 63, pp. 130–151.
______. 1994. "Laying Foods By: Gender, Dietary Decisions, and the Technology of Food Preservation in New England Households, 1750–1850." In Judith McGaw, ed., *Early American Technology: Making and Doing Things from the Colonial Era to 1850*. Chapel Hill: University of North Carolina Press for the Institute of Early American History and Culture.
Meinig, Donald W. 1969. *Imperial Texas: An Interpretive Essay in Cultural Geography*. Austin: University of Texas Press.
Nathan, Joan. 1994. *Jewish Cooking in America*. New York: Alfred A. Knopf.

The New Reliable Cookbook. 1918. Tennile, GA: J. D. Franklin Chapter, United Daughters of the Confederacy.

Nourse, Edwin G. 1918. *The Chicago Produce Market*. Boston: Houghton Mifflin.

O'Bannon, Patrick. 1994. "Inconsiderable Progress: Commercial Brewing in Philadelphia Before 1840." In Judith McGaw, ed., *Early American Technology: Making and Doing Things from the Colonial Era to 1850*. Chapel Hill: University of North Carolina Press for the Institute of Early American History and Culture.

100 Year History: 1882–1982 and Future Probe. n.d. Special issue of *Beverage World*.

Panis' Cookbook. 1978. Mercer, PA: Camp Nazareth.

Panschar, William G. 1956. *Baking in America*. Evanston, IL: Northwestern University Press.

Patteson, Charles, with Craig Emerson. 1988. *Charles Patteson's Kentucky Cooking*. New York: Harper & Row.

Perdue, Charles L. 1992. *Pigsfoot Jelly & Persimmon Beer*. Santa Fe: Ancient City Press.

The Picayune Creole Cook Book. 1901. New Orleans: Times-Picayune.

The Pillsbury Cook Book. 1914. Minneapolis: Pillsbury Flour Mills.

Pillsbury, Richard. 1990. *From Boarding House to Bistro: The American Restaurant Then and Now*. London: Unwin and Hyman.

Pillsbury, Richard, and John Florin. 1996. *Atlas of American Agriculture: The American Cornucopia*. New York: Macmillan.

Pineland Country Cooking: From the Geechee to the Hoopee. 1980. Pineland, GA: Emanual County 4-H Foundation.

Plavchan, Ronald J. 1976. *A History of Anheuser-Busch, 1852–1933*. New York: Arno Press.

Powell, William J. 1985. *Pillsbury's Best: A Company History from 1869*. Minneapolis: Pillsbury Company.

Progressive Grocer's Marketing Guidebook. Various years. Stamford, CT: Interactive Market Systems.

Purcell, L. Edward. 1995. *Immigration*. Phoenix, AZ: Oryx Press.

Purvis, Thomas. 1987. "Patterns in the Ethnic Settlement in Late Eighteenth-Century Pennsylvania." *Western Pennsylvania Historical Magazine* 70, no. 2, pp. 107–122.

Randolph, Mary. 1825. *Virginia House-Wife*. Second edition. Washington, DC: Printed for the author by Way & Gideon.

Ranhofer, Charles. [1893] 1971. *The Epicurean: A Complete Treatise of Analytical and Practical Studies on the Culinary Art*. New York: Dover.

Report of the Massachusetts Commission on the Cost of Living. 1910. Boston: State of Massachusetts.

River Road Recipes. 1963. Baton Rouge: Junior League of Baton Rouge.

Root, Waverly, and Richard de Rochemont. 1976. *Eating in America*. New York: William Morrow.

Rorer, Mrs. S. T. 1904. *World's Fair Souvenir Cook Book, Louisiana Purchase Exposition, St. Louis, 1904*. Philadelphia: Arnold.

Rundell, Maria. 1807. *A New System of Domestic Cookery: Formed upon Principles of Economy. . .* Richmond, VA: Jacob Johnson.

Sanjur, Diva. 1995. *Hispanic Foodways, Nutrition & Health*. Boston: Allyn and Bacon.

Schoonover, David, ed. 1992. *P.E.O. Cookbook*. Souvenir edition, reprint of 1908 edition. Iowa City: University of Iowa Press.

Senauer, Ben, Elain Asp, and Jean Kinsey. 1991. *Food Trends and the Changing Consumer*. St. Paul: Eagan Press.

Shortridge, Barbara, and James Shortridge. 1980. "Consumption of Fresh Vegetables in the Metropolitan United States." *Geographical Review* 79, no. 1, pp. 79–98.

______. Forthcoming. *American Regional Diets: A Regional and Ethnic Reader*. Lanham, MD: Rowan and Littlefield.

Simmons, Amelia. 1796. *American Cookery*. Introduction and updated recipes by Iris Frey. Reprinted by Green Farms, CT: Silverleaf Press, 1984.

Simoons, Frederick J. 1991. *Food in China: A Cultural and Historical Inquiry*. Boca Raton: CRC Press.

______. 1994. *Eat Not This Flesh: Food Avoidances from Prehistory to the Present*. Second edition. Madison: University of Wisconsin Press.

Simpson, J. A., and E.S.C. Weiner, preps. 1989. *The Oxford English Dictionary*. Second edition. Oxford, UK: Clarendon Press.

Smart-Grosvenor, Vertamae. 1992. *Vibration Cooking, or The Travel Notes of a Geechee Girl*. New York: Ballantine Books.

Smith, Eliza. 1758. *The Complete Housewife: Or, Accomplished Gentlewoman's Companion*. Facsimile of sixteenth edition by Kings Langley, UK: Arion House, 1983.

Sokolov, Raymond. 1991. *Why We Eat What We Eat: How the Encounter Between the New World and the Old Changed the Way Everyone on the Planet Eats*. New York: Summit Books.

Southern Living Annual Recipes. Various years. Birmingham, AL: Oxmoor House.

Standard Directory of Advertisers. Various years. New Providence, NJ: National Register.

Steelman, Virginia. 1974. *The Cultural Context of Food: A Study of Food Habits and Their Social Significance in Selected Areas of Louisiana*. Agricultural Experimental Station Bulletin 681. Baton Rouge: Louisiana State University Center for Agricultural Sciences and Rural Development.

Stokley, William B., and A. J. Stokley. 1962. *The Best Fed Nation: A 100 Year Progress Report*. New York: Newcomen Society of the United States.

Strasser, Susan. 1982. *Never Done: A History of American Housework*. New York: Pantheon Books.

The Structure of Food Manufacturing. 1966. Technical Study 8. Washington, DC: National Commission on Food Marketing.

Sturges, Lena E. 1971. *Our Best Recipes*. Birmingham, AL: Oxmoor Press.

Supermarket News. 1971. *Distribution of Food Store Sales in 288 Cities*. New York: Fairchild Publications.

Tannahill, Reay. 1973. *Food in History*. New York: Stein and Day.

Tartan, Beth. 1992. *North Carolina and Old Salem Cookery*. Revised edition. Chapel Hill: University of North Carolina Press.

Taylor, Joe Gray. 1982. *Eating, Drinking, and Visiting in the South: An Informal History*. Baton Rouge: Louisiana State University Press.

Taylor, William A. 1901. "The Influence of Refrigeration on the Fruit Industry." *Yearbook of the United States Department of Agriculture, 1900*. Washington, DC: Government Printing Office, pp. 561–580.

Thornton, Harrison J. 1933. *The History of the Quaker Oats Company*. Chicago: University of Chicago Press.

Tried n' True Recipes. 1974. Max, ND: Max Legion Auxilary, Post 241.

Truaxx, Carol. 1960. *Ladies' Home Journal Cookbook*. Garden City, NY: Doubleday.

Tyree, Marion Cabell. 1884. *Housekeeping in Old Virginia*. Louisville, KY: John P. Morton.

Van Egmond-Pannell, Dorothy. 1985. *School Foodservice*. Third edition. Westport, CT: AVI.

Wade, Louise C. 1987. *Chicago's Pride: The Stockyards, Packingtown, and Environs in the Nineteenth Century*. Urbana: University of Illinois Press.

Weaver, William. 1982. *A Quaker Woman's Cookbook: The Domestic Cookery of Elizabeth Ellicott Lea*. Philadelphia: University of Pennsylvania Press.

Weigley, Emma. 1977. *Sarah Tyson Rorer: The Nation's Instructress in Dietetics and Cookery*. Philadelphia: American Philosophical Society.

What Is America Eating? 1986. Washington, DC: National Academy Press.

What to Serve and How to Serve It: A Guide for Correct Table Usage. 1922. n.p.: International Silver Company.

Wilcox, Estelle. 1880. *Buckeye Cookery and Practical Housekeeping*. Facsimile edition published by St. Paul, MN: Minnesota Historical Society Press, 1988.

Williams, Susan. 1985. *Savory Suppers and Fashionable Feasts: Dining in Victorian America*. New York: Pantheon.

Wing, Elizabeth S., and Antoinette B. Brown. 1979. *Paleonutrition: Method and Theory in Prehistoric Foodways*. New York: Academic Press.

Winthrop, John. 1996. *The Journals of John Winthrop, 1630–1649*. Cambridge: Belknap Press of Harvard University.

Wolcott, Imogene. 1971. *The Yankee Cook Book*. Revised edition. New York: Ives Washburn.

Wonderful Ways with Soups. ca. 1952. Camden, NJ: Campbell Soup Company.

Young, Carrie, and Felicia Young. 1993. *Prairie Cooks: Glorified Rice, Three Day Buns, and Other Reminiscences.* Iowa City: University of Iowa Press.

Zelinsky, Wilbur. 1973. *The Cultural Geography of the United States*. Englewood Cliffs, NJ: Prentice-Hall.

______. 1974. "Selfward Bound? Personal Preference Patterns and the Changing Map of American Society." *Economic Geographer* 50, pp. 144–179.

______. 1985. "The Roving Palate: North America's Ethnic Restaurant Cuisines." *Geoforum* 16, pp. 51–72.

Zohary, Daniel, and Maria Hopf. 1993. *Domestication of Plants in the Old World: The Origin and Spread of Cultivated Plants in West Asia, Europe, and the Nile Valley*. Second edition. Oxford: Clarendon Press.

Index